Alfred Russel Wallace

On the Phenomena of Variation and Geographical Distribution

as illustrated by the Papilionidae of the Malayan region. Read March 17, 1864

Alfred Russel Wallace

On the Phenomena of Variation and Geographical Distribution
as illustrated by the Papilionidae of the Malayan region. Read March 17, 1864

ISBN/EAN: 9783337289065

Printed in Europe, USA, Canada, Australia, Japan

Cover: Foto ©berggeist007 / pixelio.de

More available books at **www.hansebooks.com**

I. *On the Phenomena of Variation and Geographical Distribution as illustrated by the Papilionidæ of the Malayan Region.* By ALFRED R. WALLACE, *Esq.*

(Plates I.–VIII.)

Read March 17, 1864.

WHEN the naturalist studies the habits, the structure, or the affinities of animals, it matters little to which group he especially devotes himself; all alike offer him endless materials for observation and research. But, for the purpose of investigating the phenomena of geographical distribution and of local or general variation, the several groups differ greatly in their value and importance. Some have too limited a range, others are not sufficiently varied in specific forms, while, what is of most importance, many groups have not received that amount of attention over the whole region they inhabit, which could furnish materials sufficiently approaching to completeness to enable us to arrive at any accurate conclusions as to the phenomena they present as a whole. It is in those groups which are and have long been favourites with collectors that the student of distribution and variation will find his materials the most satisfactory, from their comparative completeness.

Preeminent among such groups are the diurnal Lepidoptera or Butterflies, whose extreme beauty and endless diversity have led to their having been assiduously collected in all parts of the world, and to the numerous species and varieties having been figured in a series of magnificent works, from those of Cramer, the contemporary of Linnæus, down to the inimitable productions of our own Hewitson. But, besides their abundance, their universal distribution, and the great attention that has been paid to them, these insects have other qualities that especially adapt them to elucidate the branches of inquiry already alluded to. These are the immense development and peculiar structure of the wings, which not only vary in form more than those of any other insects, but offer on both surfaces an endless variety of pattern, colouring, and texture. The scales with which they are more or less completely covered imitate the rich hues and delicate surfaces of satin or of velvet, glitter with metallic lustre, or glow with the changeable tints of the opal. This delicately painted surface acts as a register of the minutest differences of organization,—a shade of colour, an additional streak or spot, a slight modification of outline continually recurring with the greatest regularity and fixity, while the body and all its other members exhibit no appreciable change. The wings of Butterflies, as Mr. Bates has well put it[*], "serve as a tablet on which Nature writes the story of the modifications of species;" they enable us to perceive changes that would otherwise be uncertain and difficult of observation, and exhibit to us on an enlarged scale the effects of the climatal and other physical conditions which influence more or less profoundly the organization of every living thing.

A proof that this greater sensibility to modifying causes is not imaginary may, I think, be drawn from the consideration that while the Lepidoptera as a whole are of all insects the least essentially varied in form, structure, or habits, yet in the number of their specific forms they are not much inferior to those orders which range over a much wider field of nature, and exhibit more deeply seated structural modifications. The Lepidoptera are all vegetable-feeders in their larva-state, and suckers of juices or other liquids in their perfect form. In their most widely separated groups they differ but little from a common type, and offer comparatively unimportant modifications of structure or of habits. The Coleoptera, the Diptera, or the Hymenoptera, on the other hand, present far greater and more essential variations. In either of these orders we have both vegetable- and animal-feeders, aquatic, and terrestrial, and parasitic groups. Whole families are devoted to special departments in the economy of nature. Seeds, fruits, bones, carcases, excrement, bark, have each their special and dependent insect tribes from among them; whereas the Lepidoptera are, with but few exceptions, confined to the one function of devouring the foliage of living vegetation. We might therefore anticipate that their population would be only equal to those of the sections of the other orders that have a similar uniform mode of existence; and the fact that their numbers are at all comparable with those of entire orders, so much more varied in organization and habits, is, I think, a proof that they are in general highly susceptible of specific modification.

The Papilionidæ are a family of diurnal Lepidoptera which have hitherto, by almost universal consent, held the first rank in the order; and though this position has recently been denied them, I cannot altogether acquiesce in the reasoning by which it has been proposed to degrade them to a lower rank. In Mr. Bates's most excellent paper on the Heliconidæ[†], he claims for that family the highest position, chiefly because of the imperfect structure of the fore legs, which is there carried to an extreme degree of abortion, and thus removes them further than any other family from the Hesperidæ and Heterocera, which all have perfect legs. Now it is a question whether any amount of difference which is exhibited merely in the imperfection or abortion of certain organs, can establish in the group exhibiting it a claim to a high grade of organization; still less can this be allowed when another group, along with perfection of structure in the same organs, exhibits modifications peculiar to it, together with the possession of an organ which in the remainder of the order is altogether wanting. This is, however, the position of the Papilionidæ. The perfect insects possess two characters quite peculiar to them. Mr.

---

[*] See 'The Naturalist on the Amazons,' 2nd edit. p. 412.
[†] Transactions of the Linnean Society, vol. xxiii. p. 495.

Edward Doubleday, in his 'Genera of Diurnal Lepidoptera,' says, "The Papilionidæ may be known by the apparently four-branched median nervule and the spur on the anterior tibiæ, characters found in no other family." The four-branched median nervule is a character so constant, so peculiar, and so well marked, as to enable a person to tell, at a glance at the wings only of a butterfly, whether it does or does not belong to this family; and I am not aware that any other group of Butterflies, at all comparable to this in extent and modifications of form, possesses a character in its neuration to which the same degree of certainty can be attached. The spur on the anterior tibiæ is also found in some of the Hesperidæ, and is therefore supposed to show a direct affinity between the two groups; but I do not imagine it can counterbalance the differences in neuration and in every other part of their organization. The most characteristic feature of the Papilionidæ, however, and that on which I think insufficient stress has been laid, is undoubtedly the peculiar structure of the larvæ. These all possess an extraordinary organ situated on the neck, the well-known Y-shaped tentacle, which is entirely concealed in a state of repose, but which is capable of being suddenly thrown out by the insect when alarmed. When we consider this singular apparatus, which in some species is nearly half an inch long, the arrangement of muscles for its protrusion and retraction, its perfect concealment during repose, its blood-red colour, and the suddenness with which it can be thrown out, we must, I think, be led to the conclusion that it serves as a protection to the larva by startling and frightening away some enemy when about to seize it, and is thus one of the causes which has led to the wide extension and maintained the permanence of this now dominant group. Those who believe that such peculiar structures can only have arisen by very minute successive variations, each one advantageous to its possessor, must see, in the possession of such an organ by one group, and its complete absence in every other, a proof of a very ancient origin and of very long-continued modification. And such a positive structural addition to the organization of the family, subserving an important function, seems to me alone sufficient to warrant us in considering the Papilionidæ as the most highly developed portion of the whole order, and thus in retaining it in the position which the size, strength, beauty, and general structure of the perfect insects have been generally thought to deserve.

The Papilionidæ are pretty widely distributed over the earth, but are especially abundant in the tropics, where they attain their maximum of size and beauty and the greatest variety of form and colouring. South America, North India, and the Malay Islands are the regions where these fine insects occur in the greatest profusion, and where they actually become a not unimportant feature in the scenery. In the Malay Islands in particular the giant Ornithopteræ may be frequently seen about the borders of the cultivated and forest districts, their large size, stately flight, and gorgeous colouring rendering them even more conspicuous than the generality of birds. In the shady suburbs of the town of Malacca two large and handsome Papilios (*Memnon* and *Nephelus*) are not uncommon, flapping with irregular flight along the roadway, or, in the early morning, expanding their wings to the invigorating rays of the sun. In Amboyna and other towns of the Moluccas, the magnificent *Deiphobus* and *Severus*, and occasionally even the azure-winged *Ulysses*, frequent similar situations, fluttering about the orange-trees and flower-beds, or sometimes even straying into the narrow bazaars or covered markets of the city. In Java the golden-dusted *Arjuna* may often be seen at damp places on the roadside in the mountain districts, in company with *Sarpedon*, *Bathycles*, and *Agamemnon*, and less frequently the beautiful swallow-tailed *Antiphates*. In the more luxuriant parts of these islands one can hardly take a morning's walk in the neighbourhood of a town or village without seeing three or four species of *Papilio*, and often twice that number. No less than 120 species of the family are now known to inhabit the Archipelago, and of these ninety-six were collected by myself. Twenty-nine species are found in Borneo, being the largest number in any one island, twenty-three species having been obtained by myself in the vicinity of Sarawak; Java has twenty-seven species; Celebes and the Peninsula of Malacca twenty-three each. Further east the numbers decrease, Batchian producing seventeen, and New Guinea only thirteen, though this number is certainly too small, owing to our present imperfect knowledge of that great island.

In estimating these numbers I have had the usual difficulty to encounter, of determining what to consider species and what varieties. The Malayan region, consisting of a large number of islands of generally great antiquity, possesses, compared to its actual area, a great number of distinct forms, often indeed distinguished by very slight characters, but in most cases so constant in large series of specimens, and so easily separable from each other, that I know not on what principle we can refuse to give them the name and rank of species. One of the best and most orthodox definitions is that of Pritchard, the great ethnologist, who says, that "*separate origin and distinctness of race, evinced by a constant transmission of some characteristic peculiarity of organization*," constitutes a species. Now leaving out the question of "origin," which we cannot determine, and taking only the proof of separate origin, "*the constant transmission of some characteristic peculiarity of organization*," we have a definition which will compel us to neglect altogether the *amount* of difference between any two forms, and to consider only whether the differences that present themselves are *permanent*. The rule, therefore, I have endeavoured to adopt is, that when the difference between two forms inhabiting separate areas seems quite constant, when it can be defined in words, and when it is not confined to a single peculiarity only, I have considered such forms to be species. When, however, the individuals of each locality vary among themselves, so as to cause the distinctions between the two forms to become inconsiderable and indefinite, or where the differences, though constant, are confined to one particular only, such as size, tint, or a single point of difference in marking or in outline, I class one of the forms as a variety of the other.

I find as a general rule that the constancy of species is in an inverse ratio to their range. Those which are confined to one or two islands are generally very constant. When they extend to many islands, considerable variability appears; and when they have an extensive range over a large part of the Archipelago, the amount of unstable variation is very large. These facts are explicable on Mr. Darwin's principles. When a species exists over a wide area, it must have had, and probably still possesses, great powers of dispersion. Under the different conditions of existence in various portions of its area, different variations from the type would be selected, and, were they completely isolated, would soon become distinctly modified forms; but this process is checked by the dispersive powers of the whole species, which leads to the more or less frequent intermixture of the incipient varieties, which thus become irregular and unstable. Where, however, a species has a limited range, it indicates less active powers of dispersion, and the process of modification under changed conditions is less interfered with. The species will therefore exist under one or more permanent forms according as portions of it have been isolated at a more or less remote period.

What is commonly called variation consists of several distinct phenomena which have been too often confounded. I shall proceed to consider these under the heads of—1st, simple variability; 2nd, polymorphism; 3rd, local forms; 4th, coexisting varieties; 5th, races or subspecies; and 6th, true species.

1. *Simple variability.*—Under this head I include all those cases in which the specific form is to some extent unstable. Throughout the whole range of the species, and even in the progeny of individuals, there occur continual and uncertain differences of form, analogous to that variability which is so characteristic of domestic breeds. It is impossible usefully to define any of these forms, because there are indefinite gradations to each other form. Species which possess these characteristics have always a wide range, and are more frequently the inhabitants of continents than of islands, though such cases are always exceptional, it being far more common for specific forms to be fixed within very narrow limits of variation. The only good example of this kind of variability which occurs among the Malayan Papilionidæ is in *Papilio Severus*, a species inhabiting all the islands of the Moluccas and New Guinea, and exhibiting in each of them a greater amount of individual difference than often serves to distinguish well-marked species. Almost equally remarkable are the variations exhibited in most of the species of *Ornithoptera*, which I have found in some cases to extend even to the form of the wing and the arrangement of the nervures. Closely allied, however, to these variable species are others which, though differing slightly from them, are constant and confined to limited areas. After satisfying oneself, by the examination of numerous specimens captured in their native countries, that the one set of individuals are variable and the others are not, it becomes evident that by classing all alike as varieties of one species we shall be obscuring an important fact in nature, and that the only way to exhibit that fact in its true light is to treat the invariable local form as a distinct species, even though it does not offer better distinguishing characters than do the extreme forms of the variable species. Cases of this kind are the *Ornithoptera Priamus*, which is confined to the islands of Ceram and Amboyna, and is very constant in both sexes, while the allied species inhabiting New Guinea and the Papuan Islands is exceedingly variable; and in the island of Celebes is a species closely allied to the variable *P. Severus*, but which, being exceedingly constant, I have described as a distinct species under the name of *Papilio Pertinax*.

2. *Polymorphism or dimorphism.*—By this term I understand the coexistence in the same locality of two or more distinct forms, not connected by intermediate gradations, and all of which are occasionally produced from common parents. These distinct forms generally occur in the female sex only, and the intercrossing of two of these forms does not generate an intermediate race, but reproduces the same forms in varying proportions. I believe it will be found that a considerable number of what have been classed as

varieties are really cases of *polymorphism*. Albinoism and melanism are of this character, as well as most of those cases in which well-marked varieties occur in company with the parent species, but without any intermediate forms. Under these circumstances, if the two forms breed separately, and are never reproduced from a common parent, they must be considered as distinct species, contact without intermixture being a good test of specific difference. On the other hand, intercrossing without producing an intermediate race is a test of dimorphism. I consider, therefore, that under any circumstances the term 'variety' is wrongly applied to such cases.

The Malayan Papilionidæ exhibit some very curious instances of polymorphism, some of which have been recorded as varieties, others as distinct species; and they all occur in the female sex. *Papilio Memnon*, L., is one of the most striking, as it exhibits the mixture of simple variability, local and polymorphic forms, all hitherto classed under the common title of varieties. The polymorphism is strikingly exhibited by the females, one set of which resemble the males in form, with a variable paler colouring; the others have a large spatulate tail to the hinder wings and a distinct style of colouring, which causes them closely to resemble *P. Coon*, a species of which the sexes are alike and inhabiting the same countries, but with which they have no direct affinity. The tailless females exhibit simple variability, scarcely two being found exactly alike even in the same locality. The males of the island of Borneo exhibit constant differences of the under surface, and may therefore be distinguished as a local form, while the continental specimens, as a whole, offer such large and constant differences from those of the islands that I am inclined to separate them as a distinct species—*P. Androgeus*, Cr. We have here, therefore, distinct species, local forms, polymorphism, and simple variability, which seem to me to be distinct phenomena, but which have been hitherto all classed together as varieties. I may mention that the fact of these distinct forms being one species is doubly proved. The males, the tailed and tailless females, have all been bred from a single group of the larvæ, by Messrs. Payen and Bocarmé, in Java, and I myself captured in Sumatra a male *P. Memnon*, L., and a tailed female *P. Achates*, Cr., "in copulâ."

Papilio Pammon, L., offers a somewhat similar case. The female was described by Linnæus as *P. Polytes*, and was considered to be a distinct species till Westermann bred the two from the same larvæ (see Boisduval, 'Species Générales des Lépidoptères,' p. 272). They were therefore classed as sexes of one species by Mr. Edward Doubleday, in his 'Genera of Diurnal Lepidoptera,' in 1846. Later, female specimens were received from India closely resembling the male insect, and this was held to overthrow the authority of M. Westermann's observation, and to reestablish *P. Polytes* as a distinct species; and as such it accordingly appears in the British Museum List of Papilionidæ in 1856, and in the Catalogue of the East India Museum in 1857. This discrepancy is explained by the fact of *P. Pammon* having two females, one closely resembling the male, while the other is totally different from it. A long familiarity with this insect (which, replaced by local forms or by closely allied species, occurs in every island of the Archipelago) has convinced me of the correctness of this statement; for in every place where a male allied to *P. Pammon* is found, a female resembling *P. Polytes* also occurs, and sometimes, though less frequently than on the continent, another female closely resembling the male; while not only has no male specimen of *P. Polytes* yet been found, but the female (*Polytes*) has never yet been found in localities to which the male (*Pammon*) does not extend. In this case, as in the last, distinct species, local forms, and dimorphic specimens have been confounded under the common appellation of varieties.

But, besides the true *P. Polytes*, there are several allied forms of females to be considered, namely, *P. Theseus*, Cr., *P. Melanides*, De Haan, *P. Elyros*, G. R. G., and *P. Romulus*, L. The dark female figured by Cramer as *P. Theseus* seems to be the common and perhaps the only form in Sumatra, whereas in Java, Borneo, and Timor, along with males quite identical with those of Sumatra, occur females of the *Polytes* form, although a single specimen of the true *P. Theseus*, Cr., taken at Lombock would seem to show that the two forms do occur together. In the allied species found in the Philippine Islands (*P. Alphenor*, Cr., *P. Ledebouria*, Eschsch., ♀ *P. Elyros*, G. R. G.) forms corresponding to these extremes occur along with a number of intermediate varieties, as shown by a fine series in the British Museum. We have here an indication of how dimorphism may be produced; for let the extreme Philippine forms be better suited to their conditions of existence than the intermediate connecting links, and the latter will gradually die out, leaving two distinct forms of the same insect, each adapted to some special conditions. As these conditions are sure to vary in different districts, it will often happen, as in Sumatra and Java, that the one form will predominate in the one island, the other in the adjacent one. In the island of Borneo there seems to be a third form; for *P. Melanides*, De Haan, evidently belongs to this group, and has all the chief characteristics of *P. Theseus*, with a modified coloration of the hind wings. I now come to an insect which, if I am correct, offers one of the most interesting cases of variation yet adduced. *Papilio Romulus*, L., a butterfly found over a large part of India and Ceylon, and not uncommon in collections, has always been considered a true and independent species, and no suspicions have been expressed regarding it. But a male of this form does not, I believe, exist. I have examined the fine series in the British Museum, in the East India Company's Museum, in the Hope Museum at Oxford, in Mr. Hewitson's and several other private collections, and can find nothing but females; and for this common butterfly no male partner can be found except the equally common *P. Pammon*, a species already provided with two wives, and yet to whom we shall be forced, I believe, to assign a third. On carefully examining *P. Romulus*, I find that in all essential characters,—the form and texture of the wings, the length of the antennae, the spotting of the head and thorax, and even the peculiar tints and shades with which it is ornamented,—it corresponds exactly with the other females of the *Pammon* group; and though, from the peculiar marking of the fore wings, it has at first sight a very different aspect, yet a closer examination shows that every one of its markings could be produced by slight and almost imperceptible modifications of the various allied forms. I fully believe, therefore, that I shall be correct in placing *P. Romulus* as a third Indian form of the female *P. Pammon*, corresponding to *P. Melanides*, the third form of the Malayan *P. Theseus*. I may mention here that the females of this group have a superficial resemblance to the *Polydorus* group, as shown by *P. Theseus* having been considered to be the female of *P. Antiphus*, and by *P. Romulus* being arranged next to *P. Hector*. There is no close affinity between these two groups of *Papilio*, and I am disposed to believe that we have here a case of mimicry, brought about by the same causes which Mr. Bates has so well explained in his account of Heliconidæ, and which thus led to the singular exuberance of polymorphic forms in this and allied groups of the genus *Papilio*. I shall have to devote a section of my paper to the consideration of this subject.

The third example of polymorphism I have to bring forward is *Papilio Ormenus*, Guér., which is closely allied to the well-known *P. Erechtheus*, Don., of Australia. The most common form of the female also resembles that of *P. Erechtheus*; but a totally different-looking insect was found by myself in the Aru Islands, and figured by Mr. Hewitson under the name of *P. Onesimus*, which subsequent observation has convinced me is a second form of the female of *P. Ormenus*. Comparison of this with Boisduval's description of *P. Amanga*, a specimen of which from New Guinea is in the Paris Museum, shows the latter to be a closely similar form; and two other specimens were obtained by myself, one in the island of Goram and the other in Waigiou, all evidently local modifications of the same form. In each of these localities males and ordinary females of *P. Ormenus* were also found. So far there is no evidence that these light-coloured insects are not females of a distinct species, the males of which have not been discovered. But two facts have convinced me this is not the case. At Dorey, in New Guinea, where males and ordinary females closely allied to *P. Ormenus* occur (but which seem to me worthy of being separated as a distinct species), I found one of these light-coloured females closely followed in her flight by three males, exactly in the same manner as occurs (and, I believe, occurs only) with the sexes of the same species. After watching them a considerable time, I captured the whole of them, and became satisfied that I had discovered the true relations of this anomalous form. The next year I had corroborative proof of the correctness of this opinion by the discovery in the island of Batchian of a new species allied to *P. Ormenus*, all the females of which, either seen or captured by me, were of one form, and much more closely resembling the abnormal light-coloured females of *P. Ormenus* and *P. Pandion* than the ordinary specimens of that sex. Every naturalist will, I think, agree that this is strongly confirmative of the supposition that both forms of female are of one species; and when we consider, further, that in four separate islands, in each of which I resided for several months, the two forms of female were obtained and only one form of male ever seen, and that about the same time M. Montrouzier in Woodlark Island, at the other extremity of New Guinea (where he resided several years, and must have obtained all the large Lepidoptera of the island), obtained females closely resembling mine, which, in despair at finding no appropriate partners for them, he mates with a widely different species,—it becomes, I think, sufficiently evident that this is another case of polymorphism of the same nature as those already pointed out in *P. Pammon* and *P. Memnon*. This species, however, is not only *dimorphic*, but *trimorphic*; for, in the island of Waigiou, I obtained a third female quite distinct from either of the others, and in some degree intermediate between the ordinary female and the male. The specimen is particularly interesting to those who believe, with Mr. Darwin, that extreme difference of the sexes has been gradually produced by what he terms sexual selection, since it may be supposed to exhibit one of the intermediate steps in that process which has been accidentally preserved in company with its more favoured rivals, though its extreme rarity (only one specimen having been seen to many hundreds of the other form) would indicate that it may soon become extinct.

The only other case of polymorphism in the genus *Papilio*, at all equal in interest to those I have now brought forward, occurs in America; and we have, fortunately, accurate information about it. *Papilio Turnus*, L., is common over almost the whole of temperate North America; and the female resembles the male very closely. A totally different-looking insect both in form and colour, *Papilio Glaucus*, L., inhabits the same region; and though, down to the time when Boisduval published his 'Species Général,' no connexion was supposed to exist between the two species, it is now well ascertained that *P. Glaucus* is a second female form of *P. Turnus*. In the 'Proceedings of the Entomological Society of Philadelphia,' Jan. 1863, Mr. Walsh gives a very interesting account of the distribution of this species. He tells us that in the New England States and in New York all the females are yellow, while in Illinois and further south all are black; in the intermediate region both black and yellow females occur in varying proportions. Lat. 37° is approximately the southern limit of the yellow form, and 42° the northern limit of the black form; and, to render the proof complete, both black and yellow insects have been bred from a single batch of eggs. He further states that, out of thousands of specimens, he has never seen or heard of intermediate varieties between these forms. In this interesting example we see the effects of latitude in determining the proportions in which the individuals of each form should exist. The conditions are *here* favourable to the one form, *there* to the other; but we are by no means to suppose that these conditions consist in climate alone. It is highly probable that the existence of enemies, and of competing forms of life, may be the main determining influences; and it is much to be wished that such a competent observer as Mr. Walsh would endeavour to ascertain what are the adverse causes which are most efficient in keeping down the numbers of each of these contrasted forms.

Dimorphism of this kind in the animal kingdom does not seem to have any direct relations to the reproductive powers, as Mr. Darwin has shown to be the case in plants, nor does it appear to be very general. One other case only is known to me in another family of my eastern Lepidoptera, the *Pieridæ*; and but few occur in the Lepidoptera of other countries. The spring and autumn broods of some European species differ very remarkably; and this must be considered as a phenomenon of an analogous though not of an identical nature [*]. *Araschnia prorsa*, of Central Europe, is a striking example of this alternate or seasonal dimorphism. Mr. Pascoe has pointed out two forms of the male sex in some species of Coleoptera belonging to the family Anthribidæ, in seven species of the two genera *Xenocerus* and *Mecocerus* (Proc. Ent. Soc. Lond., 1862, p. 71); and no less than six European Water-beetles, of the genus *Dytiscus*, have females of two forms, the most common having the elytra deeply sulcate, the rarer smooth as in the males. The three, and sometimes four or more, forms under which many Hymenopterous insects (especially Ants) occur must be considered as a related phenomenon, though here each form is specialized to a distinct function in the economy of the species. Among the higher animals,

[*] Among our nocturnal Lepidoptera, I am informed, many analogous cases occur; and as the whole history of many of these has been investigated by breeding successive generations from the egg, it is to be hoped that some of our British Lepidopterists will give us a connected account of all the abnormal phenomena which they present.

albinoism and melanism may, as I have already stated, be considered as analogous facts; and I met with one case of a bird, a species of Lory (*Eos fuscata*, Blyth), clearly existing under two forms, since I obtained both sexes of each from a single flock.

The fact of the two sexes of one species differing very considerably is so common, that it attracted but little attention till Mr. Darwin showed how it could in many cases be explained by what he termed sexual selection. For instance, in most polygamous animals the males fight for the possession of the females, and the victors becoming the progenitors of the succeeding generation, impress upon their male offspring their own superior size, strength, or unusually developed offensive weapons. It is thus that we can account for the spurs and the superior strength and size of the males in Gallinaceous birds, and also for the large canine tusks in the males of fruit-eating Apes. So the superior beauty of plumage and special adornments of the males of so many birds can be explained by supposing (what there are many facts to prove) that the females prefer the most beautiful and perfect-plumaged males, and that thus slight accidental variations of form and colour have been accumulated till they have produced the wonderful train of the Peacock and the gorgeous plumage of the Bird of Paradise. Both these causes have no doubt acted partially in insects, so many species possessing horns and powerful jaws in the male sex only, and still more frequently the males alone rejoicing in rich colours or sparkling lustre. But there is here another cause which has led to sexual differences, viz. a special adaptation of the sexes to diverse habits or modes of life. This is well seen in female Butterflies (which are generally weaker and of slower flight), often having colours better adapted to concealment; and in certain South American species (*Papilio torquatus*) the females, which inhabit the forests, resemble the *Æneas* group, which abound in similar localities, while the males, which frequent the sunny open river-banks, have a totally different coloration. In these cases, therefore, natural selection seems to have acted independently of sexual selection; and all such cases may be considered as examples of the simplest dimorphism, since the offspring never offer intermediate varieties between the parent forms.

The distinctive character therefore of dimorphism is this, that the union of these distinct forms does not produce intermediate varieties, but reproduces them unchanged. In simple varieties, on the other hand, as well as when distinct local forms or distinct species are crossed, the offspring never resembles either parent exactly, but is more or less intermediate between them. Dimorphism is thus seen to be a specialized result of variation, by which new physiological phenomena have been developed; the two should therefore, whenever possible, be kept separate *.

3. *Local form, or variety.*—This is the first step in the transition from variety to species.

* The phenomena of dimorphism and polymorphism may be well illustrated by supposing that a blue-eyed, flaxen-haired Saxon man had two wives, one a black-haired, red-skinned Indian squaw, the other a woolly-headed, sooty-skinned negress—and that instead of the children being mulattoes of brown or dusky tints, mingling the separate characteristics of their parents in varying degrees, all the boys should be pure Saxon boys like their father, while the girls should altogether resemble their mothers. This would be thought a sufficiently wonderful fact; yet the phenomena here brought forward as existing in the insect-world are still more extraordinary; for each mother is capable not only of producing male offspring like the father, and female like herself, but also of producing other females exactly like her fellow-wife, and altogether differing from herself. If an island could be stocked with a colony of human beings having similar physiological idiosyncrasies with *Papilio Pammon* or *Papilio Ormenus*, we should see

It occurs in species of wide range, when groups of individuals have become partially isolated in several points of its area of distribution, in each of which a characteristic form has become segregated more or less completely. Such forms are very common in all parts of the world, and have often been classed as varieties or species alternately. I restrict the term to those cases where the difference of the forms is very slight, or where the segregation is more or less imperfect. The best example in the present group is *Papilio Agamemnon*, L., a species which ranges over the greater part of tropical Asia, the whole of the Malay archipelago, and a portion of the Australian and Pacific regions. The modifications are principally of size and form, and, though slight, are tolerably constant in each locality. The steps, however, are so numerous and gradual that it would be impossible to define many of them, though the extreme forms are sufficiently distinct. *Papilio Sarpedon*, L., presents somewhat similar but less numerous variations.

4. *Coexisting variety.*—This is a somewhat doubtful case. It is when a slight but permanent and hereditary modification of form exists in company with the parent or typical form, without presenting those intermediate gradations which would constitute it a case of simple variability. It is evidently only by direct evidence of the two forms breeding separately that this can be distinguished from *dimorphism*. The difficulty occurs in *Papilio Jason*, Esp., and *P. Evemon*, Bd., which inhabit the same localities, and are almost exactly alike in form, size, and coloration, except that the latter always wants a very conspicuous red spot on the under surface, which is found not only in *P. Jason*, but in all the allied species. It is only by breeding the two insects that it can be determined whether this is a case of a coexisting variety or of *dimorphism*. In the former case, however, the difference being constant and so very conspicuous and easily defined, I see not how we could escape considering it as a distinct species. A true case of coexisting forms would, I consider, be produced, if a slight variety had become fixed as a local form, and afterwards been brought into contact with the parent species with little or no intermixture of the two; and such instances do very probably occur.

5. *Race, or subspecies.*—These are local forms completely fixed and isolated; and there is no possible test but individual opinion to determine which of them shall be considered as species and which varieties. If stability of form and "*the constant transmission of some characteristic peculiarity of organization*" is the test of a species (and I can find no other test that is more certain than individual opinion), then every one of these fixed races, confined as they almost always are to distinct and limited areas, must be regarded as a species; and as such I have in most cases treated them. The various modifications of *Papilio Ulysses*, *P. Peranthus*, *P. Codrus*, *P. Eurypilus*, *P. Helenus*, &c., are excellent examples; for while some present great and well-marked, others offer slight and inconspicuous differences, yet in all cases these differences seem equally fixed and permanent. If, therefore, we call some of these forms species, and others varieties, we introduce a purely arbitrary distinction, and shall never be able to decide where to draw the line. The races of *Papilio Ulysses*, L., for example, vary in amount of modification from the scarcely differing New Guinea form to those of Woodlark Island and New Caledonia, but

white men living with yellow, red, and black women, and their offspring always reproducing the same types; so that at the end of many generations the men would remain pure white, and the women of the same well-marked races as at the commencement.

all seem equally constant; and as most of these had already been named and described as species, I have added the New Guinea form under the name of *P. Penelope*. We thus get a little group of Ulyssine Papilios, the whole comprised within a very limited area, each one confined to a separate portion of that area, and, though differing in various amounts, each apparently constant. Few naturalists will doubt that all these may and probably have been derived from a common stock; and therefore it seems desirable that there should be a unity in our method of treating them: either call them all *varieties* or all *species*. Varieties, however, continually get overlooked; in lists of species they are often altogether unrecorded; and thus we are in danger of neglecting the interesting phenomena of variation and distribution which they present. I think it advisable, therefore, to name all such forms; and those who will not accept them as species may consider them as subspecies or races.

6. *Species.*—Species are merely those strongly marked races or local forms which, when in contact, do not intermix, and when inhabiting distinct areas are generally believed to have had a separate origin, and to be incapable of producing a fertile hybrid offspring. But as the test of hybridity cannot be applied in one case in ten thousand, and even if it could be applied, would prove nothing, since it is founded on an assumption of the very question to be decided—and as the test of separate origin is in every case inapplicable—and as, further, the test of non-intermixture is useless, except in those rare cases where the most closely allied species are found inhabiting the same area, it will be evident that we have no means whatever of distinguishing so-called "true species" from the several modes of variation here pointed out, and into which they so often pass by an insensible gradation. It is quite true that, in the great majority of cases, what we term "species" are so well marked and definite that there is no difference of opinion about them; but as the test of a true theory is, that it accounts for, or at the very least is not inconsistent with, the whole of the phenomena and apparent anomalies of the problem to be solved, it is reasonable to ask that those who deny the origin of species by variation and selection should grapple with the facts in detail, and show how the doctrine of the distinct origin and permanence of species will explain and harmonize them. It has been recently asserted by a high authority that the difficulty of limiting species is in proportion to our ignorance, and that just as groups or countries are more accurately known and studied in greater detail the limits of species become settled*. This statement has, like many other general assertions, its portion of both truth and error. There is no doubt that many uncertain species, founded on few or isolated specimens, have had their true nature determined by the study of a good series of examples: they have been thereby established as species or as varieties; and the number of times this has occurred is doubtless very great. But there are other and equally trustworthy cases in which, not single species, but whole groups have, by the study of a vast accumulation of materials, been proved to have no definite specific limits. A few of these must be adduced. In Dr. Carpenter's 'Introduction to the Study of the Foraminifera,' he states that "*there is not a single specimen of plant or animal of which the range of variation has been studied by the collocation and comparison of so large a number of specimens as have passed under the review of Messrs. Williamson, Parker, Rupert Jones, and myself, in our studies of the*

---

* See Dr. J. E. Gray "On the Species of Lemuroids," Proc. Zool. Soc. 1863, p. 134.

types of this group;" and the result of this extended comparison of specimens is stated to be, "*The range of variation is so great among the Foraminifera as to include not merely those differential characters which have been usually accounted* SPECIFIC, *but also those upon which the greater part of the* GENERA *of this group have been founded, and even in some instances those of its* ORDERS" (Foraminifera, Preface, x). Yet this same group had been divided by D'Orbigny and other authors into a number of clearly defined *families, genera,* and *species,* which these careful and conscientious researches have shown to have been almost all founded on incomplete knowledge.

Professor DeCandolle has recently given the results of an extensive review of the species of *Cupuliferæ.* He finds that the best-known species of oaks are those which produce most varieties and subvarieties, that they are often surrounded by provisional species; and, with the fullest materials at his command, two-thirds of the species he considers more or less doubtful. His general conclusion is, that "*in botany the lowest series of groups,* SUBVARIETIES, VARIETIES, *and* RACES *are very badly limited; these can be grouped into* SPECIES *a little less vaguely limited, which again can be formed into sufficiently precise* GENERA." This general conclusion is entirely objected to by the writer of the article in the 'Natural History Review,' who, however, does not deny its applicability to the particular order under discussion, while this very difference of opinion is another proof that difficulties in the determination of species do not, any more than in the higher groups, vanish with increasing materials and more accurate research.

Another striking example of the same kind is seen in the genera *Rubus* and *Rosa,* adduced by Mr. Darwin himself; for though the amplest materials exist for a knowledge of these groups, and the most careful research has been bestowed upon them, yet the various species have not thereby been accurately limited and defined so as to satisfy the majority of botanists.

Dr. Hooker seems to have found the same thing in his study of the Arctic flora. For though he has had much of the accumulated materials of his predecessors to work upon, he continually expresses himself as unable to do more than group the numerous and apparently fluctuating forms into more or less imperfectly defined species*.

Lastly, I will adduce Mr. Bates's researches on the Amazons. During eleven years he accumulated vast materials, and carefully studied the variation and distribution of insects. Yet he has shown that many species of Lepidoptera, which before offered no special difficulties, are in reality most intricately combined in a tangled web of affinities, leading by such gradual steps from the slightest and least stable variations to fixed races and well-marked species, that it is very often impossible to draw those sharp dividing-lines which it is supposed that a careful study and full materials will always enable us to do.

These few examples show, I think, that in every department of nature there occur instances of the instability of specific form, which the increase of materials aggravates

---

* In his paper on the "Distribution of Arctic Plants," Trans. Linn. Soc. xxiii. p. 310, Dr. Hooker says:—

"The most able and experienced descriptive botanists vary in their estimate of the value of the 'specific term' to a much greater extent than is generally supposed."

"I think I may safely affirm that the 'specific term' has three different standard values, all current in descriptive botany, but each more or less confined to one class of observers."

"This is no question of what is right or wrong as to the real value of the specific term, I believe each is right according to the standard he assumes as the specific."

rather than diminishes. And it must be remembered that the naturalist is rarely likely to err on the side of imputing greater indefiniteness to species than really exists. There is a completeness and satisfaction to the mind in defining and limiting and naming a species, which leads us all to do so whenever we conscientiously can, and which we know has led many collectors to reject vague intermediate forms as destroying the symmetry of their cabinets. We must therefore consider these cases of excessive variation and instability as being thoroughly well established; and to the objection that, after all, these cases are but few compared with those in which species can be limited and defined, and are therefore merely exceptions to a general rule, I reply that a true law embraces all apparent exceptions, and that to the great laws of nature there are no real exceptions—that what appear to be such are equally results of law, and are often (perhaps indeed always) those very results which are most important as revealing the true nature and action of the law. It is for such reasons that naturalists now look upon the study of *varieties* as more important than that of well-fixed species. It is in the former that we see nature still at work, in the very act of producing those wonderful modifications of form, that endless variety of colour, and that complicated harmony of relations, which gratify every sense and give occupation to every faculty of the true lover of nature.

Variation as specially influenced by Locality.

The phenomena of variation as influenced by locality have not hitherto received much attention. Botanists, it is true, are acquainted with the influences of climate, altitude, and other physical conditions in modifying the forms and external characteristics of plants; but I am not aware that any peculiar influence has been traced to locality, independent of climate. Almost the only case I can find recorded is mentioned in that repertory of natural-history facts, 'The Origin of Species,' viz. that herbaceous groups have a tendency to become arboreal in islands. In the animal world, I cannot find that any facts have been pointed out as showing the special influence of locality in giving a peculiar *facies* to the several disconnected species that inhabit it. What I have to adduce on this matter will therefore, I hope, possess some interest and novelty.

On examining the closely allied species, local forms, and varieties distributed over the Indian and Malayan regions, I find that larger or smaller districts, or even single islands, give a special character to the majority of their Papilionidæ. For instance: 1. The species of the Indian region (Sumatra, Java, and Borneo) are almost invariably smaller than the allied species inhabiting Celebes and the Moluccas; 2. The species of New Guinea and Australia are also, though in a less degree, smaller than the nearest species or varieties of the Moluccas; 3. In the Moluccas themselves the species of Amboyna are the largest; 4. The species of Celebes equal or even surpass in size those of Amboyna; 5. The species and varieties of Celebes possess a striking character in the form of the anterior wings, different from that of the allied species and varieties of all the surrounding islands; 6. Tailed species in India or the Indian region become tailless as they spread eastward through the archipelago.

Having preserved the finest and largest specimens of Butterflies in my own collection, and having always taken for comparison the largest specimens of the same sex, I believe that the tables I now give are sufficiently exact. The differences of expanse of wings are in most cases very great, and are much more conspicuous in the specimens themselves than on paper. It will be seen that no less than fourteen Papilionidæ inhabiting Celebes and the Moluccas are from one-third to one-half greater in extent of wing than the allied species representing them in Java, Sumatra, and Borneo. Six species inhabiting Amboyna are larger than the closely allied forms of the northern Moluccas and New Guinea by about one-sixth. These include almost every case in which closely allied species can be compared.

PAPILIONIDÆ.

Species of the Moluccas and Celebes (large).	Expanse. inches.	Closely allied species of Java and the Indian region (small).	Expanse. inches.
Ornithoptera Helena (Amboyna)	7·6	O. Pompeus	5·8
		O. Amphrisius	6·0
Papilio Macedon (Celebes)	5·8	P. Peranthus	3·8
P. Philippus (Moluccas)	4·8		
P. Blumei (Celebes)	5·4	P. Brama	4·0
P. Alphenor (Celebes)	4·8	P. Theseus	3·6
P. Gigon (Celebes)	5·4	P. Demolion	4·0
P. Deucalion (Celebes)	4·6	P. Macareus	3·7
P. Agamemnon, var. (Celebes)	4·4	P. Agamemnon, var.	3·8
P. Eurypilus (Moluccas)	4·0	P. Jason	3·4
P. Telephus (Celebes)	4·3		
P. Ægisthus (Moluccas)	4·4	P. Rama	3·2
P. Miletus (Celebes)	4·4	P. Sarpedon	3·8
P. Androcles (Celebes)	4·8	P. Antiphates	3·7
P. Polyphontes (Celebes)	4·6	P. Diphilus	3·9
Leptocircus Curtius (Celebes)	2·0	L. Meges	1·8
Species inhabiting Amboyna (large).		Allied species of New Guinea and the North Moluccas (smaller).	
Papilio Ulysses	6·4	P. Penelope	5·2
		P. Telegonus	4·0
P. Polydorus	4·9	P. Leodamas	4·0
P. Deiphobus	6·8	P. Deiphontes	5·8
P. Gambrisius	6·4	P. Ormenus	5·6
		P. Tydeus	6·0
P. Codrus	5·4	P. Codrus, var. *papuensis*	4·4
Ornithoptera Priamus, ♂	8·0	Orn. Poseidon, ♂	7·0

The differences of form are equally clear.

Papilio Pammon everywhere on the continent is tailed in both sexes. In Java, Sumatra, and Borneo, the closely allied *P. Theseus* has a very short tail, or tooth only, in the male, while in the females the tail is retained. Further east, in Celebes and the South Moluccas, the hardly separable *P. Alphenor* has quite lost the tail in the male, while the female retains it, but in a narrower and less spatulate form. A little further, in Gilolo, *P. Nicanor* has completely lost the tail in both sexes.

Papilio Agamemnon exhibits a somewhat similar series of changes. In India it is always tailed; in the greater part of the archipelago it has a very short tail; while far east, in New Guinea and the adjacent islands, the tail has almost entirely disappeared.

In the *Polydorus*-group two species, *P. Antiphus* and *P. Diphilus*, inhabiting India and the Indian region, are tailed, while the two which take their place in the Moluccas, New Guinea, and Australia, *P. Polydorus* and *P. Leodamus*, are destitute of tail, the species furthest east having lost this ornament the most completely.

Western species, tailed.	Eastern species (closely allied), less tailed.
Papilio Pammon (India) . . . . . tailed.	P. Thesus (islands) . . . . very short tail.
P. Agamemnon, var. (India) . . . tailed.	P. Agamemnon, var. (islands) not tailed.
P. Antiphus (India, Java) . . . . tailed.	P. Polydorus (Moluccas) . not tailed.
P. Diphilus (India, Java) . . . . tailed.	P. Leodamus (New Guinea) . not tailed.

The most conspicuous instance of local modification of form, however, is exhibited in the island of Celebes, which in this respect, as in some others, stands alone and isolated in the whole archipelago. Almost every species of *Papilio* inhabiting Celebes has the wings of a peculiar shape, which distinguishes them at a glance from the allied species of every other island. This peculiarity consists, first, in the upper wings being generally more elongate and falcate; and secondly, in the costa or anterior margin being much more curved, and in most instances exhibiting near the base an abrupt bend or elbow, which in some species is very conspicuous. This peculiarity is visible, not only when the Celebesian species are compared with their small-sized allies of Java and Borneo, but also, and in an almost equal degree, when the large forms of Amboyna and the Moluccas are the objects of comparison, showing that this is quite a distinct phenomena from the difference of size which has just been pointed out.

In the following Table I have arranged the chief Papilios of Celebes in the order in which they exhibit this characteristic form most prominently. (See Plate VIII.)

Papilios of Celebes, having the wings falcate or with abruptly curved costa.	Closely allied Papilios of the surrounding islands, with less falcate wings and slightly curved costa.
1. P. Gigon, n. s.	P. Demolion (Java).
2. P. Telephus, n. s.	P. Jason (Sumatra).
3. P. Miletus, n. s.	P. Sarpedon (Moluccas, Java).
4. P. Agamemnon, var.	P. Agamemnon, var. (Borneo).
5. P. Macedon, n. s.	P. Peranthus (Java).
6. P. Ascalaphus.	P. Deiphontes, n. s. (Gilolo).
7. P. Hecuba, n. s.	P. Helenus (Java).
8. P. Blumei.	P. Brama (Sumatra).
9. P. Androcles.	P. Antiphates (Borneo).
10. P. Rhesus.	P. Aristæus (Moluccas).
11. P. Theseus, var., ♂ .	P. Thesus, ♀ (Java).
12. P. Codrus, var.	P. Codrus (Moluccas).
13. P. Encelades.	P. Leucothoë (Malacca).

It thus appears that every species of *Papilio* exhibits this peculiar form in a greater or less degree, except one, *P. Polyphontes*, Bd., allied to *P. Diphilus* of India and *P. Polydorus* of the Moluccas. This fact I shall recur to again, as I think it helps us to understand something of the causes that may have brought about the phenomenon we are considering. Neither do the genera *Ornithoptera* and *Leptocircus* exhibit any traces of this peculiar form. In several other families of Butterflies this characteristic form reappears in a few species. In the Pieridæ the following species exhibit it distinctly:—

1. Eronia tritæa	. . . .	compared with	Eronia Valeria (Java).	
2. Iphias Glaucippe, var.	. .	,,	,,	Iphias Glaucippe (Java).
3. Pieris Zebuda	. . . . .	,,	,,	Pieris Descombesi (India).
4. P. Zarinda	. . . . .	,,	,,	P. Nero (Malacca).
5. P., n. s.	. . . . .	,,	,,	P. Hyparete (Java).
6. P. Hombronii	. . ⎫ have the same form, but are isolated species.			
7. P. Ithome	. . . ⎭			
8. P. Eperia, *Bd.*	. . . . .	compared with	P. Coronis (Java).	
9. P. Polisma	. . . .	,,	,,	P., n. s. (Malacca).
10. Terias, n. s.	. . . . .	,,	,,	P. Tilaha (Java).

The other species of *Terias*, one or two *Pieris*, and the genus *Callidryas* do not exhibit any perceptible change of form.

In the other families there are but few similar examples. The following are all that I can find in my collection:—

Cethosia Æole	. . . .	compared with	Cethosia Biblis (Java).
Junonia, n. s.	. . . .	,, ,,	Junonia Polynice (Borneo).
Limenitis Limire	. . . . . .	,, ,,	Limenitis Procris (Java).
Cynthia Arsinoë, var.	. . .	,, ,,	Cynthia Arsinoë (Java, Sum., Born.).

All these belong to the family of the Nymphalidæ. Many other genera of this family, as *Diadema*, *Adolias*, *Charaxes*, and *Cyrestis*, as well as the entire families of the Danaidæ, Satyridæ, Lycænidæ, and Hesperidæ, present no examples of this peculiar form of the upper wing in the Celebesian species.

The facts now brought forward seem to me of the highest interest. We see that almost all the species in two important families of the Lepidoptera (Papilionidæ and Pieridæ) acquire, in a single island, a characteristic modification of form distinguishing them from the allied species and varieties of all the surrounding islands. In other equally extensive families no such change occurs, except in one or two isolated species. However we may account for these phenomena, or whether we may be quite unable to account for them, they furnish, in my opinion, a strong corroborative testimony in favour of the doctrine of the origin of species by successive small variations; for we have here slight varieties, local races, and undoubted species, all modified in exactly the same manner, indicating plainly a common cause producing identical results. On the generally received theory of the original distinctness and permanence of species, we are met by this difficulty: one portion of these curiously modified forms are admitted to have been produced by variation and some natural action of local conditions; whilst the other portion, differing from the former only in degree, and connected with them by insensible gradations, are said to have possessed this peculiarity of form at their first creation, or to have derived it from unknown causes of a totally distinct nature. Is not the *à priori* evidence in favour of the assumption of an identity of the causes that have produced such similar results? and have we not a right to call upon our opponents for some proofs of their own doctrine, and for an explanation of its difficulties, instead of their assuming that they are right, and laying upon us the burthen of disproof?

Let us now see if the facts in question do not themselves furnish some clue to their own explanation. Mr. Bates has shown that certain groups of butterflies have a defence against insectivorous animals, independent of swiftness of motion. These are generally very abundant, slow, and weak fliers, and are more or less the objects of mimicry by other groups, which thus gain an advantage in a freedom from persecution similar to that enjoyed by those they resemble. Now the only Papilios which have not in Celebes acquired the peculiar form of wing belong to a group which is imitated both by other species of *Papilio* and by Moths of the genus *Epicopeia*, West. This group is of weak and slow flight; and we may therefore fairly conclude that it possesses some means of defence (probably in a peculiar odour or taste) which saves it from attack. Now the arched costa and falcate form of wing is generally supposed to give increased powers of flight, or, as seems to me more probable, greater facility in making sudden turnings, and thus baffling a pursuer. But the members of the *Polydorus*-group (to which belongs the only unchanged Celebesian *Papilio*), being already guarded against attack, have no need of this increased power of wing; and "natural selection" would therefore have no tendency to produce it. The whole family of Danaidæ are in the same position: they are slow and weak fliers; yet they abound in species and individuals, and are the objects of mimicry. The Satyridæ have also probably a means of protection—perhaps their keeping always near the ground and their generally obscure colours; while the Lycænidæ and Hesperidæ may find security in their small size and rapid motions. In the extensive family of the Nymphalidæ, however, we find that several of the larger species, of comparatively feeble structure, have their wings modified (*Cethosia*, *Limenitis*, *Junonia*, *Cynthia*), while the large-bodied powerful species, which have all an excessively rapid flight, have exactly the same form of wing in Celebes as in the other islands. On the whole, therefore, we may say that all the butterflies of rather large size, conspicuous colours, and not very swift flight have been affected in the manner described, while the smaller-sized and obscure groups, as well as those which are the objects of mimicry, and also those of exceedingly swift flight, have remained unaffected.

It would thus appear as if there must be (or once have been) in the island of Celebes, some peculiar enemy to these larger-sized butterflies which does not exist, or is less abundant, in the surrounding islands. Increased powers of flight, or rapidity of turning, was advantageous in baffling this enemy; and the peculiar form of wing necessary to give this would be readily acquired by the action of "natural selection" on the slight variations of form that are continually occurring. Such an enemy one would naturally suppose to be an insectivorous bird; but it is a remarkable fact that most of the genera of Fly-catchers of Borneo and Java on the one side (*Muscipeta*, *Philentoma*), and of the Moluccas on the other (*Monarcha*, *Rhipidura*), are almost entirely absent from Celebes. Their place seems to be supplied by the Caterpillar-catchers (*Graucalus*, *Campephaga*), of which six or seven species are known from Celebes and are very numerous in individuals. We have no positive evidence that these birds pursue butterflies on the wing, but it is highly probable that they do so when other food is scarce*. However this may be, the fauna of Celebes is undoubtedly highly peculiar in every department of which we have

* Mr. Bates has suggested that the larger Dragon-flies (*Eshna*, &c.) prey upon butterflies; but I did not notice that they were more abundant in Celebes than elsewhere.

any knowledge; and though we may not be able to trace it satisfactorily, there can, I think, be little doubt that the singular modification in the wings of so many of the butterflies of that island is an effect of that complicated action and reaction of all living things upon each other in the struggle for existence, which continually tends to readjust disturbed relations, and to bring every species into harmony with the varying conditions of the surrounding universe.

But even the conjectural explanation now given fails us in the other cases of local modification. Why the species of the western islands should be smaller than those further east, —why those of Amboyna should exceed in size those of Gilolo and New Guinea—why the tailed species of India should begin to lose that appendage in the islands, and retain no trace of it on the borders of the Pacific, are questions which we cannot at present attempt to answer. That they depend, however, on some general principle is certain, because analogous facts have been observed in other parts of the world. Mr. Bates informs me that, in three distinct groups, Papilios which on the Upper Amazon and in most other parts of South America have spotless upper wings obtain pale or white spots at Pará and on the Lower Amazon; and also that the Æneus-group of Papilios never have tails in the equatorial regions and the Amazons valley, but gradually acquire tails in many cases as they range towards the northern or southern tropic. Even in Europe we have somewhat similar facts; for the species and varieties of butterflies peculiar to the island of Sardinia are generally smaller and more deeply coloured than those of the mainland, and *Papilio Hospiton* has lost the tail, which is a prominent feature of the closely allied *P. Machaon*.

Facts of a similar nature to those now brought forward would no doubt be found to occur in other groups of insects, were local faunas carefully studied in relation to those of the surrounding countries; and they seem to indicate that climate and other physical causes have, in some cases, a very powerful effect in modifying specific form, and thus directly aid in producing the endless variety of nature.

I may state that I can adduce facts perfectly analogous to these from other families of Lepidoptera, especially the Danaidæ; but as the greater part of the species are still undescribed, I can only now assert that similar phenomena do occur there.

Mimicry.

I need scarcely say that I entirely agree with Mr. Bates's explanation of the causes which have led to one group of insects mimicking another (Trans. Linn. Soc. vol. xxiii. p. 495). I have, therefore, only now to adduce such illustrations of this curious phenomenon as are furnished by the Eastern Papilionidæ, and to show their bearing upon the phenomena of variation already mentioned. As in America, so in the Old World, species of Danaidæ are the objects which the other families most often imitate. But, besides these, some genera of Morphidæ and one section of the genus *Papilio* are also less frequently copied. Many species of *Papilio* mimic other species of these three groups so closely that they are undistinguishable when on the wing; and in every case the pairs which resemble each other inhabit the same locality.

The following list exhibits the most important and best-marked cases of mimicry which occur among the Papilionidæ of the Malayan region and India:—

Mimickers*.	Species mimicked.	Common habitat.
	DANAIDÆ.	
1. Papilio paradoxa, Zink., ♂	Euplœa Midamus, Cr., ♂	} Sumatra, &c.
—— ——, ♀	—— ——, ♀	
2. —— ——, West.	E. Rhadamanthus	Sumatra, &c.
3. P. Caunus,	E., sp.	Borneo.
4. P. Thule, Wall.	Danais sobrina, Bd.	New Guinea.
5. P. Macareus, Godt.	D. Aglaia, Cr.	Malacca, Java.
6. P. Agestor, G. R. G.	D. Tytia, G. R. G.	Northern India.
7. P. idæoides, Hewits.	Hestia Leuconoë, Erichs.	Philippines.
8. P. Delessertii, Guér.	Hestia, sp.	Penang.
	MORPHIDÆ.	
9. P. Paudion, Wall., ♀	Drusilla bioculata, Guér.	New Guinea.
	PAPILIO (POLYDORUS- and COON-groups).	
10. P. Pammon, L. (Romulus, L.), ♀	Papilio Hector, L.	India.
11. P. Theseus, Cr., var., ♀	P. Antiphus, Fab.	Sumatra, Borneo.
12. P. Theseus, Cr., var., ♀	P. Diphilus, Esp.	Sumatra, Java.
13. P. Memnon, var. Achates, ♀	P. Coon, Fab.	Sumatra.
14. P. Androgeus, var. Achates, ♀	P. Doubledayi, Wall.	Northern India.
15. P. Œnomaus, God., ♀	P. Liris, God.	Timor.

We have therefore fifteen species or marked varieties of *Papilio* which so closely resemble species of other groups in their respective localities, that it is not possible to impute the resemblance to accident. The first two in the list (*Papilio paradoxa* and *P. Caunus*) are so exactly like *Euplœa Midamus* and *E. Rhadamanthus* on the wing, that, although they fly very slowly, I was quite unable to distinguish them. The first is a very interesting case, because the male and female differ considerably, and each mimics the corresponding sex of the *Euplœa*. A new species of *Papilio* which I discovered in New Guinea resembles *Danais sobrina*, Bd., from the same country, just as *Papilio Macareus* resembles *Danais Aglaia* in Malacca, and (according to Dr. Horsfield's figure) still more closely in Java. The Indian *Papilio Agestor* closely imitates *Danais Tytia*, which has quite a different style of colouring from the preceding; and the extraordinary *Papilio idæoides* from the Philippine Islands must, when on the wing, perfectly resemble the *Hestia Leuconoë* of the same region, as also does the *P. Delessertii*, Guér., imitate an undescribed species of *Hestia* from Penang. Now in every one of these cases the Papilios are very scarce, while the Danaidæ which they resemble are exceedingly abundant—most of them swarming so as to be a positive nuisance to the collecting entomologist by continually hovering before him when he is in search of newer and more varied captures. Every garden, every roadside, the suburbs of every village are full of them, indicating

---

* The terms "mimicry" and "mimickers" have been objected to on the ground that they imply voluntary action on the part of the insects. This appears to me of little importance compared with the advantages of convenience, flexibility, and expressiveness which they undoubtedly possess, especially as the whole theory propounded by the originator of the term in this sense excludes all idea of voluntary action. The only approximately synonymous words, not implying will, are *resemblance*, *similarity*, and *likeness*; and it is evident that none of these can be applied intelligibly under the variety of forms required, and to which Mr. Bates's expression so readily lends itself in the terms *mimic*, *mimickers*, *mimicry*, *mimicked*. Add to this the inconvenience of changing a term which, from the interest and wide discussion of the subject, must be already very generally understood, and I think it will be admitted that nothing would be gained by altering it, even if a better word were pointed out, which has not yet been done.

very clearly that their life is an easy one, and that they are free from persecution by the foes which keep down the population of less favoured races. This superabundant population has been shown by Mr. Bates to be a general characteristic of all American groups and species which are objects of mimicry; and it is interesting to find his observations confirmed by examples on the other side of the globe.

The remarkable genus *Drusilla*, a group of pale-coloured butterflies, more or less adorned with ocellate spots, is also the object of mimicry by three distinct genera (*Melanitis*, *Hyantis*, and *Papilio*). These insects, like the *Danaidæ*, are abundant in individuals, have a very weak and slow flight, and do not seek concealment, or appear to have any means of protection from insectivorous creatures. It is natural to conclude, therefore, that they have some hidden property which saves them from attack; and it is easy to see that when any other insects, by what we call accidental variation, come more or less remotely to resemble them, the latter will share to some extent in their immunity. An extraordinary dimorphic form of a female *Papilio* has come to resemble the Drusillas sufficiently to be taken for one of that group at a little distance; and it is curious that I captured one of these Papilios in the Aru Islands hovering along the ground, and settling on it occasionally, just as it is the habit of the Drusillas to do. The resemblance in this case is only general; but this form of *Papilio* varies much, and there is therefore material for natural selection to act upon so as ultimately to produce a copy as exact as in the other cases.

The eastern Papilios allied to *Polydorus Coon* and *P. Philoxenus*, form a natural section of the genus resembling, in many respects, the *Eneus*-group of South America, which they may be said to represent in the East. Like them, they are forest insects, have a low and weak flight, and in their favourite localities are rather abundant in individuals; and like them, too, they are the objects of mimicry. We may conclude, therefore, that they possess some hidden means of protection, which makes it useful to other insects to be mistaken for them.

The Papilios which resemble them belong to a very distinct section of the genus, in which the sexes differ greatly; and it is those females only which differ most from the males, and which have already been alluded to as exhibiting instances of dimorphism, which resemble species of the other group.

The resemblance of *P. Romulus* to *P. Hector* is, in some specimens, very considerable, and has led to the two species being placed to follow each other in the British Museum Catalogues and by Mr. E. Doubleday. I have shown, however, that *P. Romulus* is probably a dimorphic form of the female *P. Pammon*, and belongs to a distinct section of the genus*.

The next pair, *P. Theseus*, Cr., and *P. Antiphus*, Fab., have been united as one species both by De Haan and in the British Museum Catalogues. The ordinary variety of *P. Theseus* found in Java almost as nearly resembles *P. Diphilus*, Esp., of the same country. The most interesting case, however, is the extreme female form of *P. Memnon* (*P. Achates*, Cr.)†, which has acquired the general form and markings of *P. Coon*, an insect which differs from the ordinary male *P. Memnon*, as much as any two species differ which can be chosen in this extensive and highly varied genus; and, as if to show that this resemblance is not accidental, but is the result of law, when in India we find a species closely allied to

* See Plate II. fig. 6.   † See Plate I. fig. 4.

P. Coon, but with red instead of yellow spots (*P. Doubledayi*, Wall.), the corresponding variety of *P. Androgeus* (*P. Achates*, Cram., 182, A, B.) has acquired exactly the same peculiarity of having red spots instead of yellow. Lastly, in the island of Timor, the female of *P. Œnomaus* (a species allied to *P. Memnon*) resembles so closely *P. Liris* (one of the *Polydorus*-group), that the two, which were often seen flying together, could only be distinguished by a minute comparison after being captured.

The last six cases of mimicry are especially instructive, because they seem to indicate one of the processes by which dimorphic forms have been produced. When, as in these cases, one sex differs much from the other, and varies greatly itself, it may happen that occasionally individual variations will occur having a distant resemblance to groups which are the objects of mimicry, and which it is therefore advantageous to resemble. Such a variety will have a better chance of preservation; the individuals possessing it will be multiplied; and their accidental likeness to the favoured group will be rendered permanent by hereditary transmission, and, each successive variation which increases the resemblance being preserved, and all variations departing from the favoured type having less chance of preservation, there will in time result those singular cases of two or more isolated and fixed forms bound together by that intimate relationship which constitutes them the sexes of a single species. The reason why the females are more subject to this kind of modification than the males is, probably, that their slower flight, when laden with eggs, and their exposure to attack while in the act of depositing their eggs upon leaves, render it especially advantageous for them to have some additional protection. This they at once obtain by acquiring a resemblance to other species which, from whatever cause, enjoy a comparative immunity from persecution.

This summary of the more interesting phenomena of variation presented by the eastern Papilionidæ is, I think, sufficient to substantiate my position, that the Lepidoptera are a group that offer especial facilities for such inquiries; and it will also show that they have undergone an amount of special adaptive modification rarely equalled among the the more highly organized animals. And, among the Lepidoptera, the great and preeminently tropical families of Papilionidæ and Danaidæ seem to be those in which complicated adaptations to the surrounding organic and inorganic universe have been most completely developed, offering in this respect a striking analogy to the equally extraordinary, though totally different, adaptations which present themselves in the *Orchideæ*, the only family of plants in which mimicry of other organisms appears to play any important part, and the only one in which striking cases of polymorphism occur; for such we must consider to be the male, female, and hermaphrodite forms of *Catasetum tridentatum*, which differ so greatly in form and structure that they were long considered to belong to three distinct genera.

Arrangement and Geographical Distribution of the Malayan Papilionidæ.

Although the species of Papilionidæ inhabiting the Malayan region are very numerous, they all belong to three out of the nine genera into which the family is divided. One of the remaining genera (*Eurycus*) is restricted to Australia, and another (*Teinopalpus*) to the Himalayan Mountains, while no less than four (*Parnassius*, *Doritis*, *Thais*, and *Sericinus*) are confined to Southern Europe and to the mountain-ranges of the Palæarctic region.

The genera *Ornithoptera* and *Leptocircus* are highly characteristic of Malayan entomology, but are uniform in character and of small extent. The genus *Papilio*, on the other hand, presents a great variety of forms, and is so richly represented in the Malay islands, that more than one-fourth of all the known species are found there. It becomes necessary, therefore, to divide this genus into natural groups before we can successfully study its geographical distribution.

Owing principally to Dr. Horsfield's observations in Java, we are acquainted with a considerable number of the larvæ of Papilios; and these furnish good characters for the primary division of the genus into natural groups. The manner in which the hinder wings are plaited or folded back at the abdominal margin, the size of the anal valves, the structure of the antennæ, and the form of the wings are also of much service, as well as the character of the flight and the style of coloration. Using these characters, I divide the Malayan Papilios into four sections, and seventeen groups, as follows:—

Genus ORNITHOPTERA.

    a. *Priamus*-group. Black and green.   b. *Pompeus*-group. Black and yellow.
    c. *Brookeanus*-group.

Genus PAPILIO.

A. Larvæ short, thick, with numerous fleshy tubercles; purplish.
    a. *Nox*-group. Abdominal fold in ♂ very large; anal valves small, but swollen; antennæ moderate; wings entire, or tailed: includes the Indian *Philoxenus*-group.
    b. *Coon*-group. Abdominal fold in ♂ small; anal valves small, but swollen; antennæ moderate; wings tailed.
    c. *Polydorus*-group. Abdominal fold in ♂ small, or none; anal valves small or obsolete, hairy; wings tailed or entire.

B. Larvæ with third segment swollen, transversely or obliquely banded; pupa much bent. Imago with abdominal margin in ♂ plaited, but not reflexed; body weak; antennæ long; wings much dilated, often tailed.   d. *Ulysses*-group.
    e. *Peranthus*-group. } *Protenor*-group (Indian) is somewhat intermediate between these, and is
    f. *Memnon*-group. } nearest to the *Nox*-group.
    g. *Helenus*-group.   h. *Erectheus*-group.
    i. *Pammon*-group.   k. *Demolion*-group.

C. Larvæ subcylindrical, variously coloured. Imago with abdominal margin in ♂ plaited, but not reflexed; body weak; antennæ short, with a thick curved club; wings entire.
    l. *Erithonius*-group. Sexes alike, larva and pupa something like those of *P. Demolion*.
    m. *Paradoxa*-group. Sexes different.
    n. *Dissimilis*-group. Sexes alike; larva bright-coloured: pupa straight, cylindric.

D. Larvæ elongate, attenuate behind, and often bifid, with lateral and oblique pale stripes, green. Imago with the abdominal margin in ♂ reflexed, woolly or hairy within; anal valves small, hairy; antennæ short, stout; body stout.
    o. *Macareus*-group. Hind wings entire.
    p. *Antiphates*-group. Hind wings much tailed (swallow-tails).
    q. *Eurypylus*-group. Hind wings elongate or tailed.

Genus LEPTOCIRCUS.

making, in all, **twenty distinct** groups of Malayan Papilionidæ.

The first section of the genus *Papilio* (A) comprises insects which, though differing considerably in **structure**, have **much** general resemblance. They all have a weak, low flight, frequent the most luxuriant forest-districts, seem to love the shade, and are the objects of mimicry by other Papilios.

Section B consists of weak-bodied, large-winged insects, with an irregular wavering flight, and which, when resting on foliage, often expand the wings, which the species of the other sections rarely or never do. They are the most conspicuous and striking of of eastern Butterflies.

Section C consists of much weaker and slower-flying insects, often resembling in their flight, as well as in their colours, species of Danaidæ.

Section D contains the strongest-bodied and most swift-flying of the genus. They love sunlight, and frequent the borders of streams and the edges of puddles, where they gather together in swarms consisting of several species, greedily sucking up the moisture, and, when disturbed, circling round in the air, or flying high and with great strength and rapidity.

In the following Table I have arranged all the Malayan Papilionidæ in what appears to me their most natural succession, and have exhibited their distribution in twenty-one columns of localities, extending from the Malay peninsula, on the north-west, to Woodlark Island, near New Guinea, on the south-east. The double line divides the Indo-Malayan from the Austro-Malayan region; and those islands which form natural zoological groups are connected by brackets.

Table showing the Distribution of the Malayan Papilionidæ.

		Indo-Malayan Region.				Austro-Malayan Region.													
	Ornithoptera.	Malacca	Sumatra	Borneo	Java	Philippines	Celebes	Sooloo	Timor	Gilolo	Batchian	Ceram	Goram	Ke Islands	Aru Islands	Mysol	Waigiou	New Guinea	New Ireland	Woodlark Isl.
	a. *Pompeus*-group.																			
1	Priamus, *L.*				1						1									
2	Poseidon, *Db.*														1	1	1	1	1	1
3	Cræsus, *Feld.*					1	1													
4	Tithonus, *De Haan*															1				
5	Urvillianus, *Godt.*																		1	
	b. *Pompeus*-group.																			
6	Remus, *Cr.*					1		1												
7	Helena, *L.*					1				1	1							1		
8	Lydius, *Wall.*																			
9	Pompeus, *Cr.*	1	1	1	1	1														
10	Nephereus, *G.R.G.*					1														
11	Magellanus, *Feld.*					1														
12	Criton, *Feld.*								1	1										
13	Plato, *Wall.*							1												
14	Haliphron, *Bd.*						1													
15	Amphrisius, *Cr.*	1	1	1																
	c. *Brookeana*-group.																			
16	Brookeana, *Wall.*		1	1																
	Papilio.																			
A.	a. *Nox*-group.																			
17	Nox, *Sw.*		1		1															
18	Noctis, *Hew.*			1																
19	Erebus, *Wall.*	1		1																

OF THE MALAYAN REGION.

Table showing the Distribution of the Malayan Papilionidae (continued).

	Papilio	Indo-Malayan Region.					Austro-Malayan Region.												
		Malacca	Sumatra	Borneo	Java	Philippines	Celebes	Lombock	Timor	Gilolo	Batchian	Bouru	Ceram	Matabello	Goram	Ké Island	Aru Islands	Mysol	Waigiou	New Guinea	New Ireland	Woodlark Isl.
	a. *Nox-group (continued).*																		
20	Varuna, *White*	1																	
21	Semperi, *Feld.*					1													
	b. *Coon-group.*																		
22	Neptunus, *Guér.*	1		1															
23	Coon, *Fab.*		1	1	1														
	c. *Polydorus-group.*																		
24	Polydorus, *L.*						1	1	1			1	1					1	
25	Leodamas, *Wall.*																	1	
26	Diphilus, *Esper*	1			1	1													
27	Antiphus, *Fab.*		1	1	1	1		1											
28	Polyphontes, *Bd.*						1			1	1								
29	Annæ, *Feld.*					1													
30	Liris, *Godt.*								1										
	d. *Ulysses-group.*																		
31	Ulysses, *L.*									1									
32	Penelope, *Wall.*																1	1	1
33	Telegonus, *Feld.*						1	1											
34	Telemachus, *Mont.*																			1
	e. *Peranthus-group.*																		
35	Peranthus, *Fab.*			1		1													
36	Pericles, *Wall.*							1											
37	Philippus, *Wall.*									1	1								
38	Macedon, *Wall.*					1													
39	Brama, *Guér.*	1	1																
40	Dædalus, *Feld.*				1														
41	Blumei, *Bd.*						1												
42	Arjuna, *Horsf.*	1	1	1															
	f. *Memnon-group.*																		
43	Memnon, *L.*	1	1	1		1													
44	Androgeus, *Cr.*	1																	
45	Lampsacus, *Bd.*			1															
46	Priapus, *Bd.*	1	1	1															
47	Enalthion, *Hübn.*				1														
48	Deiphontes, *Wall.*									1	1								
49	Deiphobus, *L.*											1	1						
50	Ascalaphus, *Bd.*						1												
51	Ænomaus, *Godt.*							1											
	g. *Helenus-group.*																		
52	Severus, *Cr.*									1	1	1	1				1		
53	Pertinax, *Wall.*					1													
54	Albinus, *Wall.*																		1
55	Phæstus, *Bd.*																		1
56	Helenus, *L.*	1		1															
57	Hecuba, *Wall.*					1													
58	Iswara, *White*	1		1															
59	Hystaspes, *Feld.*					1													
60	Araspes, *Feld.*					1													
61	Nephelus, *Bd.*	1	1	1															
	h. *Pammon-group.*																		
62	Pammon, *L.*	1																	
63	Theseus, *Cr.*		1	1	1		1	1											

Table showing the Distribution of the Malayan Papilionidæ (continued).

	Papilio. h. Priamus-group (continued).	Indo-Malayan Region.					Austro-Malayan Region.												
		Malacca	Sumatra	Borneo	Java	Philippines	Celebes	Lombock	Timor	Goilolo	Batchian	Ceram	Banda	Goram	Ké Island	Aru Islands	Mysol	Waigiou	New Guinea	New Ireland	Woodlark Id.
64	Alphenor, *Cr.*					1	1					1	1								
65	Nicanor, *Wall.*									1	1										
66	Hipponous, *Feld.*					1															
67	Ambrax, *Bd.*																1		1		
68	Ambracia, *Wall.*																		1		
69	Epirus, *Wall.*																1				
70	Dunali, *Monte.*																				1
	i. *Erechtheus*-group.																				
71	Ormenus, *Guér.*														1	1	1		1		
72	Pandion, *Wall.*																1		1		
73	Tydeus, *Feld.*									1	1										
74	Adrastus, *Wall.*													1							
75	Gambrisius, *Cr.*											1	1								
76	Anaphytrion, *Cr.*						1														
77	Euchenor, *Guér.*														1	1		1	1		
78	Godartii, *Monte.*																				1
	k. *Demolion*-group.																				
79	Demolion, *Cr.*	1	1	1	1																
80	Gigon, *Wall.*						1														
	l. *Erithonius*-group.																				
81	Erithonius, *Cr.*	1				1		1							1						
	m. *Paradoxa*-group.																				
82	Paradoxa, *Zink.*		1	1	1																
83	Ænigma, *Wall.*	1	1	1																	
84	Caunus, *Westw.*		1	1																	
85	Antiphus, *Westw.*				1																
86	Hewitsonii, *Westw.*		1																		
	n. *Dissimilis*-group.																				
87	Echidna, *De Haan.*							1													
88	Pacyphates, *Westw.*					1															
	o. *Macareus*-group.																				
89	Vejovis, *Hew.*					1															
90	Encelades, *Bd.*					1															
91	Demolion, *Bd.*					1															
92	Idæoides, *Hew.*				1																
93	Delessertii, *Guér.*	1																			
94	Dehaanii, *Wall.*	1		1	1																
95	Leucothoë, *Westw.*	1																			
96	Macareus, *God.*	1		1	1																
97	Stratocles, *Feld.*					1															
98	Thule, *Wall.*																		1	1	
	p. *Antiphates*-group.																				
99	Antiphates, *Cr.*	1	1	1	1																
100	Euphrates, *Feld.*					1															
101	Androcles, *Bd.*						1														
102	Dorcus, *De Haan.*						1														
103	Rheus, *Bd.*						1														
104	Aristæus, *Cr.*									1		1									
105	Parmatus, *G. R. G.*															1		1			

OF THE MALAYAN REGION.

Table showing the Distribution of the Malayan Papilionidæ (continued).

	Papilio (continued).	Indo-Malayan Region.				Austro-Malayan Region.												
		Malacca	Sumatra	Borneo	Java	Philippines	Celebes	Lombock	Timor	Gilolo	Batchian	Bouru	Ceram	Banda	Goram	Ke Island	Aru Islands	Mysol	Waigiou	New Guinea	New Ireland	Woodlark Id.
	q. Eurypylus-group.																					
106	Codrus, Cr.					1				1	1	1	1				1		1			
107	Melanthus, Feld.				1																	
108	Empedocles, Feld.			1	1																	
109	Payeni, Bd.			1																		
110	Sarpedon, L.	1	1	1	1					1	1	1	1				1			1		
111	Miletus, Wall.					1																
112	Wallacei, Hew.									1							1					
113	Bathycles, Zink.	1		1	1																	
114	Eurypylus, L.						1	1	1	1										1		
115	Jason, Esp.	1	1	1	1																	
116	Telephus, Wall.					1																
117	Ægistus, L.									1	1		1				1	1				
118	Agamemnon, L.	1	1	1	1	1	1	1	1	1	1	1	1				1	1	1	1	1	
119	Rama, Feld.	1	1																			
	(? Arycles, Bd.)																					
	Leptocircus.																					
120	Meges, Zink.	1			1																	
121	Curtius, Wall.					1																
122	Decius, Feld.					1																
123	Curius, Fab.	1			1																	
	Totals :—																					
	Ornithoptera	2	2	3	2	2	4	1	1	3	2	1	3		1	1	1	1	1	3	1	1
	Papilio	22	19	26	23	17	19	5	7	11	15	9	13	1	2	4	12	4	8	11		4
	Leptocircus	1			2	1	1															
	Species in each island	25	21	29	27	20	24	6	8	14	17	10	16	1	3	5	13	5	9	14	1	5
	Total	45					20	24	12				27					27				
		Sixty-one, Indo-Malayan Region.								Seventy-two, Austro-Malayan Region.												

The exceeding richness of the Malayan region in these fine insects is seen by comparing the number of species found in the different tropical regions of the earth. From all Africa only 33 species of *Papilio* are known; but as several are still undescribed in collections, we may raise their number to about 40. In all tropical Asia there are at present described only 65 species, and I have seen in collections but two or three which have not yet been named. In South America, south of Panama, there are 120 species, or about the same number as I make in the Malayan region; but the area of the two countries is very different; for while South America (even excluding Patagonia) contains 5,000,000 square miles, a line encircling the whole of the Malayan islands would only include an area of 2,700,000 square miles, of which the land-area would be about 1,000,000 square miles. This superior richness is partly real and partly apparent. The breaking up of a district into small isolated portions, as in an archipelago, seems highly favourable to the segregation and perpetuation of local peculiarities in certain groups; so that a species which on a continent might have a wide range, and whose local forms, if any, would be so connected together that it would be impossible to separate them, may become by isolation reduced to a number of such clearly defined and constant forms that we are obliged to count them as species. From this point of view, therefore, the superior number of Malayan species may be considered as apparent only. Its true superiority is shown, on the other hand, by the possession of three genera and twenty groups of Papilionidæ against a single genus and eight groups in South America, and also by the much greater average size of the Malayan species. In most other families, however, the reverse is the case, the South American *Nymphalidæ*, *Satyridæ*, and *Erycinidæ* far surpassing those of the East in number, variety, and beauty.

The following list, exhibiting the range and distribution of each group, will enable us to study more easily their internal and external relations.

Range of the Groups of Malayan Papilionidæ.

Ornithoptera.
1. *Priamus*-group. Moluccas to Woodlark Island.
2. *Pompeus*-group. Himalayas to New Guinea (Celebes, maximum).
3. *Brookeana*-group. Sumatra and Borneo.

Papilio.
4. *Nox*-group. North India, Java, and Philippines.
5. *Coon*-group. North India to Java.
6. *Polydorus*-group. India to New Guinea and Pacific.
7. *Ulysses*-group. Celebes to New Caledonia.
8. *Peranthus*-group. India to Timor and Moluccas (India, max.).
9. *Memnon*-group. India to Timor and Moluccas (Java, max.).
10. *Helenus*-group. Africa and India to New Guinea.
11. *Pammon*-group. India to Pacific and Australia.
12. *Erechtheus*-group. Celebes to Australia.
13. *Demolion*-group. India to Celebes.
14. *Erithonius*-group. Africa, India, Australia.
15. *Paradoxa*-group. India to Java (Borneo, max.).
16. *Dissimilis*-group. India to Timor (India, max.).
17. *Macareus*-group. India to New Guinea.
18. *Antiphates*-group. Widely distributed.
19. *Eurypylus*-group. India to Australia.

Leptocircus.
20. *Leptocircus*-group. India to Celebes.

This Table shows the great affinity of the Malayan with the Indian Papilionidæ, only three out of the nineteen groups ranging beyond, into Africa, Europe, or America. The limitation of groups to the Indo-Malayan or Austro-Malayan divisions of the archipelago, which is so well marked in the higher animals (see 'Journal of Linnean Society,' vol. iv. 172, and 'Journal of the Royal Geographical Society,' 1863, p. 230), is much less conspicuous in insects, but is shown in some degree by the Papilionidæ. The following groups are either almost or entirely restricted to one portion of the Archipelago:—

Indo-Malayan Region.	Austro-Malayan Region.
Nox-group.	*Priamus*-group.
Coon-group.	*Ulysses*-group.
Macareus-group (nearly).	*Erechtheus*-group.
Paradoxa-group.	
Dissimilis-group (nearly).	
Brookeanus-group.	
LEPTOCIRCUS (genus).	

The remaining groups, which range over the whole archipelago, are, in many cases, insects of very powerful flight, or they frequent open places and the sea-beach, and are thus more likely to get blown from island to island. The fact that three such characteristic groups as those of *Priamus*, *Ulysses*, and *Erechtheus* are strictly limited to the Australian region of the archipelago, while five other groups are with equal strictness confined to the Indian region, is a strong corroboration of that division which has been founded almost entirely on the distribution of Mammalia and Birds.

If the various Malayan islands have undergone recent changes of level, and if any of them have been more closely united within the period of existing species than they are now, we may expect to find indications of such changes in community of species between islands now widely separated; while those islands which have long remained isolated would have had time to acquire peculiar forms by a slow and natural process of modification.

An examination of the relations of the species of the adjacent islands will thus enable us to correct opinions formed from a mere consideration of their relative positions. For example, looking at a map of the archipelago, it is almost impossible to avoid the idea that Java and Sumatra have been recently united; their present proximity is so great, and they have such an obvious resemblance in their volcanic structure. Yet there can be little doubt that this opinion is erroneous, and that Sumatra has had a more recent and more intimate connexion with Borneo than it has had with Java. This is strikingly shown by the mammals of these islands—very few of the species of Java and Sumatra being identical, while a considerable number are common to Sumatra and Borneo. The birds show a somewhat similar relationship; and we shall find that the group of insects we are now treating of tells exactly the same tale. Thus:—

Sumatra . . . . . 21 sp.
Borneo . . . . . . 29 sp. } 20 sp. common to both islands;

Sumatra . . . . . 21 sp.
Java . . . . 27 sp. } 11 sp. common to both islands;

Borneo . . . . . . 29 sp.
Java . . . . 27 sp. } 20 sp. common to both islands;

showing that both Sumatra and Java have a much closer relationship to Borneo than they have each other—a most singular and interesting result when we consider the wide separation of Borneo from them both, and its very different structure. The evidence furnished by a single group of insects would have had but little weight on a point of such magnitude if standing alone; but coming as it does to confirm deductions drawn from whole classes of the higher animals, it must be admitted to have considerable value.

We may determine in a similar manner the relations of the different Papuan Islands to New Guinea. Of thirteen species of Papilionidæ obtained in the Aru Islands, five were also found in New Guinea, and eight not. Of nine species obtained at Waigiou, five were New Guinea, and four not. The five species found at Mysol were all New Guinea species. Mysol, therefore, has closer relations to New Guinea than the other islands; and this is corroborated by the distribution of the birds, of which I will only now give one instance. The Paradise Bird found in Mysol is the common New Guinea species, while the Aru Islands and Waigiou have each a species peculiar to themselves.

The large island of Borneo, which contains more species of Papilionidæ than any other in the archipelago, has nevertheless only two peculiar to itself; and it is quite possible, and even probable, that one of these may be found in Sumatra or Java. The last-named island has also two species peculiar to it; Sumatra has not one, and the peninsula of Malacca only one. The identity of species is even greater than in birds or in most other groups of insects, and points very strongly to a recent connexion of the whole with each other and the continent. But when we pass to the next island (Celebes), separated from them by a strait not wider than that which divides them from each other, we have a striking contrast; for with a total number of species less than either Borneo or Java, no less than eighteen are absolutely restricted to it. Further east, the large islands of Ceram and New Guinea have only three species peculiar to each, and Timor has five. We shall have to look, not to single islands, but to whole groups, in order to obtain an amount of individuality comparable with that of Celebes. For example, the extensive group comprising the large islands of Java, Borneo, and Sumatra, with the peninsula of Malacca, possessing altogether 45 species, has about 21, or less than half, peculiar to it; the numerous group of the Philippines possess 21 species, of which 16 are peculiar; the seven chief islands of the Moluccas have 27, of which 12 are peculiar; and the whole of the Papuan Islands, with an equal number of species, have 17 peculiar. Comparable with the most isolated of these groups is Celebes, with its 24 species, of which the large proportion of 18 are peculiar. We see, therefore, that the opinion I have already expressed, in the papers before quoted, of the high degree of isolation and the remarkable distinctive features of this interesting island, is fully borne out by the examination of this conspicuous family of insects. A single straggling island, with a few small satellites, it is zoologically of equal importance with extensive groups of islands many times as large as itself; and standing in the very centre of the archipelago, surrounded on every side with islets connecting it with the larger groups, and which seem to afford the greatest facilities for the migration and intercommunication of their respective productions, it yet stands out conspicuous with a character of its own in every department of nature, and presents peculiarities which are, I believe, without a parallel in any similar locality on the globe.

Briefly to summarize these peculiarities, Celebes possesses three genera of mammals (out of the very small number which inhabit it) which are of singular and isolated forms, viz., *Cynopithecus*, a tailless Ape allied to the Baboons; *Anoa*, a straight-horned

Antelope of obscure affinities, but quite unlike anything else in the whole archipelago or in India; and *Babirusa*, an altogether abnormal wild Pig. With a rather limited bird population, Celebes has an immense preponderance of species confined to it, and has also five remarkable genera (*Meropogon, Streptocitta, Enodes, Scissirostrum,* and *Megocephalon*) entirely restricted to its narrow limits, as well as two others (*Prioniturus* and *Basilornis*) which only range to a single island beyond it.

Mr. Smith's elaborate tables of the distribution of Malayan Hymenoptera (see 'Proc. Linn. Soc.' Zool. vol. vii.) show that, out of the large number of 301 species collected in Celebes, 190 (or nearly two-thirds) were absolutely restricted to it, although Borneo, on one side, and the various islands of the Moluccas on the other, were equally well explored by me; and no less than twelve of the genera are not found in any other island of the archipelago. I have just shown in the present paper that, in the Papilionidæ, it has far more species of its own than any other island, and a greater proportion of peculiar species than many of the large groups of islands in the archipelago—and that it gives to a large number of the species and varieties which inhabit it, 1st, an increase of size, and, 2nd, a peculiar modification in the form of the wings, which stamp upon the most dissimilar insects a mark distinctive of their common birth-place.

What, I would ask, are we to do with phenomena such as these? Are we to rest content with that very simple, but at the same time very unsatisfying explanation, that all these insects and other animals were created exactly *as* they are, and originally placed exactly *where* they are, by the inscrutable will of their Creator, and that we have nothing to do but to register the facts and wonder? Was this single island selected for a fantastic display of creative power, merely to excite a child-like and unreasoning admiration? Is all this appearance of gradual modification by the action of natural causes—a modification the successive steps of which we can almost trace—all delusive? Is this harmony between the most diverse groups, all presenting analogous phenomena, and indicating a dependence upon physical changes of which we have independent evidence, all false testimony? If I could think so, the study of nature would have lost for me its greatest charm. I should feel as would the geologist, if you could convince him that his interpretation of the earth's past history was all a delusion—that strata were never formed in the primeval ocean, and that the fossils he so carefully collects and studies are no true record of a former living world, but were all created just as they now are, and in the rocks where he now finds them.

I must here express my own belief that none of these phenomena, however apparently isolated or insignificant, can ever stand alone—that not the wing of a butterfly can change in form, or vary in colour, except in harmony with, and as a part of, the grand march of nature. I believe, therefore, that all the curious phenomena I have just recapitulated are immediately dependent on the last series of changes, organic and inorganic, in these regions; and as the phenomena presented by the island of Celebes differ from those of all the surrounding islands, it can, I conceive, only be because the past history of Celebes has been to some extent unique and different from theirs. We must have much more evidence to determine exactly in what that difference has consisted. At present, I only see my way clear to one deduction, viz., that Celebes represents one of the oldest parts of the archipelago, that it has been formerly more completely isolated both from India and from Australia than it is now, and that, amid all the mutations it has undergone, a relic or substratum of the fauna and flora of some more ancient land has been here preserved to us.

It is only since my return home, and since I have been able to compare the productions of Celebes side by side with those of the surrounding islands, that I have been fully impressed with their peculiarity, and the great interest that attaches to them. The plants and the reptiles are still almost unknown; and it is to be hoped that some enterprising naturalist may soon devote himself to their study. The geology of the country would also be well worth exploring, and its recent fossils would be of especial interest as elucidating the changes which have led to its present anomalous condition. This island stands, as it were, upon the boundary-line between two worlds. On one side is that ancient Australian fauna which preserves to the present day the facies of an early geological epoch; on the other is the rich and varied fauna of Asia, which seems to contain, in every class and order, the most perfect and highly organized animals. Celebes has relations to both, yet strictly belongs to neither; it possesses characteristics which are altogether its own; and I am convinced that no single island upon the globe would so well repay a careful and detailed research into its past and present history.

In the following catalogue of the Malayan species of Papilionidæ I have included those from Woodlark Island, collected by M. Montrouzier, as that island comes fairly within the limits of the archipelago; while I exclude New Caledonia as belonging more to the Australian and Pacific fauna. I have given full particulars of the variation of the several species, and have described all new species, forms, varieties, and undescribed sexes. The distribution of each species is noted chiefly from my own observations[*]. As the full synonymy and references to almost every work on Lepidoptera are given in the British Museum List of Papilionidæ, I have not thought it necessary to do more than to refer to a good figure and description in well-known works; and I have quoted Boisduval's ' Species Général des Lépidoptères ' throughout. In all cases, however, where I have myself corrected the synonymy, or determined sexes which had been before improperly located, I have given much fuller references.

I have found it necessary to describe and name twenty new species, and to separate six or seven more which have been hitherto considered as varieties or sexes of other species. I have also described and separated twenty-five local forms or races, and twenty polymorphous forms or sexes, as well as several simple varieties. On the other hand, I have reduced fourteen species, which figure in some of our latest lists, to the rank of sexes or local or polymorphic forms of other species. For convenience of reference, I add a list of these, with a reference to the page where will be found the reasons for not adopting them.

Ornithoptera Pronomus, *G. R. Gray,* = O. Poseidon, *Db.* (var.), p. 36.
Ornithoptera Archideus, *G. R. Gray,* = O. Poseidon, *Db.* (var.), p. 36.
Ornithoptera Euphorion, *G. R. Gray,* = O. Poseidon, *Db.* (♀ var.), p. 36.
Ornithoptera Amphimedon, *Cr.,* = O. Helena, *L.* ♀, p. 38.
Papilio Hegemon, *G. R. Gray,* = P. Polyphontes, *Bd.*, p. 43.

[*] Species collected by myself have (Wall.) after the localities where I have found them.

Papilio Melanides, *De Haan*,   = P. Theseus, *Fab.* (♀ form), p. 54.
Papilio Romulus, *Cr.*,   = P. Pammon, *L.* (♀ form), p. 52.
Papilio Rumanzovia, *Eschsch.*,   = P. Emalthion, *Hubn.* (♀ form), p. 48.
Papilio Polytes, *L.*,   = P. Pammon, *L.*, ♀, p. 51.
Papilio Orophanes, *Bd.*,   = P. Ambrax, *Bd.*, ♀, p. 54.
Papilio Elyros, *G. R. Gray*,   = P. Alphenor, *Cr.* (♀ form), p. 55.
Papilio Amanga, *Bd.*,   = P. Ormenus, *Guér.* (♀ form), p. 55.
Papilio Onesimus, *Hewits.*,   = P. Ormenus, *Guér.* (♀ form), p. 55.
Papilio Drusius, *Cr.*,   = P. Gambrisius, *Cr.*, ♀, p. 58.

As the arrangement of the species of *Papilio* which I have adopted in this paper is somewhat new, and I hope will be found to be more natural than those which have been previously used, I here add lists of the Indian and Australian species arranged in the same manner. Those already included in my Malayan list will be indicated thus, (Mal.), and printed in *italics*.

List of the PAPILIONIDÆ of the Indian Region.

1. Teinopalpus imperialis, *Hope.*
2. Ornithoptera Darsius, *G. R. G.* (Ceylon).
3. —— Rhadamanthus, *Bd.*
4. —— *Pompeus*, Cr. (Mal.).
5. —— *Amphrisius*, Cr. (Mal.).

Papilio (Sect. A).
Nox group.

6. Papilio *Varuna*, White (Mal.).
7. —— Aidoneus, *Db.*
8. —— Philoxenus, *G. R. G.*
9. —— Polyeuctes, *Db.*
10. —— Dasarada, *Moore.*
11. —— Ravana, *Moore.*
12. —— Minereus, *G. R. G.*
13. —— Icarius, *Westw.*
14. —— Bootes, *Westw.*
15. —— Janaka, *Moore.*

Coön group.
16. Papilio Doubledayi, *Wall.*

Polydorus group.
17. Papilio Jophon, *G. R. G.* (Ceylon).
18. —— *Diphilus*, Esp. (Mal.).
19. —— Aleinous, *Klug.*
20. —— Mencius, *Feld.*
21. —— Hector, *L.*

Papilio (Sect. B).
Protenor group.
22. Papilio Protenor, *Cr.*

23. Papilio Elphenor, *Db.*
24. —— Rhetenor, *Westw.*
25. —— Sakontala, *Hewits.*

Peranthus group.
26. Papilio Crino, *Fab.* (Ceylon).
27. —— Bianor, *Cr.*
28. —— Polyctor, *Bd.*
29. —— Ganesa, *Db.*
30. —— Arcturus, *Westw.*
31. —— Paris, *L.*
32. —— Palinurus, *Fab.* ?
33. —— Krishna, *Moore.*

Memnon group.
34. Papilio *Androgeus*, Cr. (Mal.).
35. —— Polymnestor, *Cr.* (Ceylon).
36. —— Demetrius, *Cr.*

Helenus group.
37. Papilio *Helenus*, L. (Mal.).
38. —— Chaon, *Westw.*
39. —— Castor, *Westw.*
40. —— *Nephelus*, Bd. (Mal.).

Pammon group.
41. Papilio *Pammon*, L. (Mal.).

Demolion group.
42. Papilio *Demolion*, Cr. (Mal.).

Papilio (Sect. C).
Erithonius group.
43. Papilio *Erithonius*, Cr. (Mal.).

Paradoxa group.
44. Papilio *Telearchus*, *Hewits*.
45. —— Slateri, *Hewits*.

Dissimilis group.
46. Papilio dissimilis, *L.*
47. —— Panope, *L.*
48. —— Lacedæmon, *Fab.*
49. —— Pollux, *Westw.*

Papilio (Sect. D).
Macareus group.
50. Papilio *Macareus*, God. (Mal.).
51. —— *Leucothoë*, Westw. (Mal.).
52. —— Megarus, *Westw.*
53. —— Agestor, *G. R. G.*
54. —— Epytides, *Hewits.*
55. —— Xenocles, *Db.*

Antiphates group.
56. Papilio *Antiphates*, Cr. (Mal.).
57. —— Agetes, *Westw.*
58. —— Anticrates, *Db.*
59. —— Orestes, *Fab.*
60. —— Alebion, *G. R. G.*
61. —— Glycerion, *G. R. G.*

Eurypylus group.
62. Papilio Gyas, *Westw.*
63. —— Evan, *Db.*
64. —— Cloanthus, *Westw.*
65. —— *Sarpedon*, L. (Mal.).
66. —— Chiron, *Wall.*
67. —— Jason, Esp. (Mal.).
68. —— Agamemnon, L. (Mal.).
69. —— Rama, Feld. (Mal.).

4. Chinese species.
61. Indian species.
4. Ceylon species.

List of the PAPILIONIDÆ of the Australian Region.

Ornithoptera (*Priamus* group).
1. Ornithoptera *Poseidon*, Db. (Mal.).
2. —— Richmondia, *G. R. G.*

Papilio (Sect. A).
Polydorus group.
3. Papilio *Leodamas*, Wall. (Mal.).
4. —— *Liris*, Godt. (Mal.).
5. —— Godartianus, *Bd.* (Pacific Islands).

Papilio (Sect. B).
Helenus group.
6. Papilio Capaneus, *Westw.*
7. —— Ilioneus, *Don.*

Ulysses group.
8. Papilio Ulyssinus, *Westw.*
9. —— Montrouzieri, *Bd.* (New Caledonia).

Pammon group.
10. Papilio Canopus, *Westw.*

Erectheus group.
11. Papilio Erectheus, *Don.*
12. —— Amyntor, *Bd.* (New Caledonia).

Papilio (Sect. C).
Erithonius group.
13. Papilio *Erithonius*, Cr. (Mal.).

Anactor group.
14. Papilio Anactor, *McL.*

Papilio (Sect. D).
Antiphates group.
15. Papilio Læosthenes, *Db.*
16. —— Pertinax, G. R. G. (Mal.).

Eurypylus group.
17. Papilio *Sarpedon*, L. (Mal.).
18. —— Gelon, *Bd.* (New Caledonia).
19. —— Lycaon, *Westw.*
20. —— Macleayanus, *Leach.*
21. —— Scottianus, Feld. (Ash Islands).
22. Euryeus Cressida, *Fab.*

6. Pacific Islands.
16. Australia.

Catalogue of Malayan PAPILIONIDÆ.

ORNITHOPTERA (Boisd.).

The characters in the *larva* and *pupa* which have been supposed to distinguish this genus from PAPILIO are erroneous, or at least do not exist in all the species. My own observations on *O. Poseidon* show that the *larva* has no "external sheath" to the thoracic tentacles, and that the suspending thread passes round the pupa, and is not "fastened on each side to a silky tubercle." There remain therefore only the characters of the perfect insect, the most important of which are the anal valves in the male. These are very large, ovate or rounded, coriaceous, and not hairy, and are furnished with projecting points or spines (sometimes very conspicuous) which serve to attach the male more firmly to the female *in copulâ*. In several species I have observed, these points or hooks were buried in the protruded anal gland of the female, and thus effectually prevented the great weight of the insects causing them to separate upon suddenly taking flight. The great strength and size of these insects, the thick texture of their wings, their long curved and stout antennæ, their peculiar form, colour, and distribution, are the only other characters that separate them from *Papilio*. Though these may not perhaps be technically sufficient, I think it advisable and convenient to retain a genus so well known and long established.

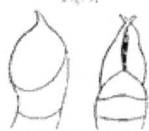

Fig. 1.

Anal valves of *O. Amphrisius*.

Ornithoptera is pre-eminently a Malayan genus, seventeen species inhabiting the archipelago, one (*Rhadamanthus*, Bd.) India and China, one (*Darsius*, G. R. Gray) peculiar to Ceylon, one (*Richmondia*, G. R. Gray) North Australia. *O. Victoriæ*, G. R. Gray, from some island east of New Guinea, should probably be included in the Malayan list; and *Æacus*, Felder, from an unknown locality. The following are the well-established Malayan species.

a. *Priamus* group.

1. ORNITHOPTERA PRIAMUS, Linnæus.

♂. *Papilio Priamus*, L.; Cram. Pap. Ex. t. 23. f. A, B; Godart, Enc. Méth. ix. p. 25. *O. Priamus*, Bd. Sp. Gén. Lép. p. 173.

♀. *P. Panthous*, L.; Cram. Pap. Ex. t. 123. f. A, t. 124. f. A.

This may be at once distinguished from all the allied species with which it has been often confounded—in the *male*, by the more rounded and deeply scalloped hind wings, with larger black spots and a broader border, the upper wings with no green on the median nervure or its branches, and the sooty patch extending only to the second median nervule; in the *female*, by the very constant and peculiar light olive-brown colour, the absence of any spots in the discoidal cell of the upper wings, and the broad shallow scallops of the hinder margin.

Hab. Amboyna and Ceram, probably also Bouru (*Wall.*).

2. ORNITHOPTERA POSEIDON, Doubleday.

O. Poseidon, Db. Ann. of Nat. Hist. xvi. p. 173; Westwood, Cat. of Orient. Ent. pl. 11, 14.

The numerous specimens of *Ornithoptera* which I obtained in various parts of New Guinea and the adjacent islands show so much instability of form, colouring, and even of neuration, no two individuals being exactly alike, that I am obliged to include them all in one variable species, to which I believe must also be referred *O. Pronomus*, G. R. Gray, from Cape York, *O. Euphorion*, G. R. Gray, from North Australia, *O. Archideus*, G. R. Gray (ex Boisd.), erroneously said to be from Celebes, and *O. Boisducalii*, Montrouzier, from Woodlark Island.

Var. *a*, Aru Islands (*Wall.*). *O. Arruana*, Feld. Lep. Frag. p. 24.

Individuals from this locality differ in the arrangement of the nervures; in some the third subcostal nervure of the upper wings branches from the same point with the upper disco-cellular, in others considerably beyond it; the points from which the subcostal nervures branch also vary. The amount of green colour on the median nervure and its branches varies. In some specimens there is a spot at the anal angle of lower wings beneath, agreeing with *O. Pronomus*, G. R. Gray; but this is generally wanting.

Var. *b*, Dorey, Salwatty, south-west coast of New Guinea (*Wall.*).

These agree very closely with *O. Poseidon*, as figured by Westwood; they differ individually in the same manner as the last, and also in the length of the lower disco-cellular nervure on the under wings. They have generally no golden spots beneath the wings. They vary also in the outline of the under wings, the outer and anal angles being more acute in some specimens than in others. Some have the under wings of a uniform green entirely without spots, while others have a range of black spots more or less fully developed.

Var. *c*, Waigiou (*Wall.*). *Archideus*, G. R. Gray, ♀.

This agrees with the last; but the male is of a more delicate green than any of the others, and has less of that colour on the median veins. On the under side there are no golden spots. The whole surface has a golden tinge, and the central portion of the lower wings is tinged with amber-brown.

The females of all the above vary extremely, much more even than the males, and from the same locality two specimens are rarely alike. The discoidal cell is in some specimens more than half occupied by a whitish patch, while in others there are only a few small spots. One of my specimens from Salwatty almost exactly agrees with that figured by Westwood (Cat. of Or. Ent. pl. 14) as from Cape York. One of the Waigiou specimens is the same as *Archideus*, G. R. G., figured by Boisduval (Voy. de l'Astrolabe, t. 4. f. 1, 2); and another, from New Guinea, differs very little from *Euphorion*, G. R. G. (Brit. Mus. Cat. Lep. pt. 1. pl. 2. f. 3), from North Australia.

From these facts I am led to conclude that we have here a variable form spread over an extensive area, and kept variable by the continual intercrossing of individuals, which would otherwise segregate into distinct and sharply defined races. The same area is inhabited by many species of birds common to all parts of it; and just as the birds of Ceram and Amboyna are almost all distinct species from those of New Guinea, so do we find those islands inhabited by the *Ornithoptera Priamus*, a well-marked and constant species, readily distinguishable in either sex from the inconstant **forms of New Guinea** proper. The same parallel holds in North Australia. Many New Guinea species of birds extend, with very slight variation, to the country about Cape York; but when we reach the Moreton Bay district all these have disappeared, and we find only true Australian species. So the variable forms of *O. Poseidon* reach North Australia and Cape York, while in the Moreton Bay district we find the comparatively well-marked species *O. Richmondia*. Similar causes, whether geographical or climatal, have thus produced an analogous distribution in these widely separated groups of animals.

3. ORNITHOPTERA CROESUS, Felder.

O. Croesus, Feld. Wien. Ent. Monats., Dec. 1859. *O. Croesus*, G. R. Gray, Proc. Zool. Soc. 1859, p. 424.

Hab. Batchian (Moluccas) (*Wall.*).

Local form, *a.—Male*: has the orange colour of the upper surface of a much deeper fiery-red hue; on the under surface, the black spots of the lower wings are nearer the margin, and the yellow spots below them are entirely absent; there is also a green line between the subcostal nervure and the margin; on the under surface of the fore wings the green patch in the discoidal cell extends to its base, and is reflexed in a narrow line along its upper edge.

Female: differs still more from that sex in *O. Croesus*; the white markings on all the wings are so large as almost to fill up the spaces between the veins, the lower part of the discoidal cell in both upper and under wings being also occupied with a whitish patch; the range of spots occupying the posterior margin are of a dusky yellow colour.

Hab. Ternate (♂), Gilolo (♀) (*Wall.*).

This well-marked local form is no doubt peculiar to Gilolo and the small adjacent islands, as the original species is to Batchian.

I was three months in the island of Batchian before I obtained a specimen of this fine insect, which I had seen once or twice only flying high in the air. I at length came upon it flying about a beautiful cinchonaceous shrub with white bracts and yellow flowers (*Mussaenda*, sp.); and having cleared a path round about, I visited the place every morning on my way to the forest, and once or twice a week had the satisfaction of capturing a fine male specimen of *O. Croesus*. The females were more plentiful and more easily caught. I afterwards sent out one of my men with a net every day to look after this insect only. He would stay out all day long, wandering up a broad rocky torrent, where the males flew up and down occasionally or settled on the rocks which just appeared above the water. He generally brought me one, and sometimes even two or three specimens; and thus, with those that I myself captured at the flowers, I secured a fine series of this richly coloured species.

4. ORNITHOPTERA TITHONUS, De Haan.

O. Tithonus, De Haan, Verh. Nat. Gesch. Ned. t. 1. f. 1.

Hab. S.W. Coast of New Guinea (*Leyden Museum*).

This remarkable species must be very rare, as I never saw it in any part of the New

Guinea district that I visited; nor was it seen during the exploration, a few years ago, by a Dutch steamer which visited the part of the coast where the only specimen known was said to have been obtained.

5. ORNITHOPTERA URVILLIANA, Guérin.

Papilio Urvilliana, Guér. Voy. de la Coquille, Lép. t. 13. f. 1, 2, ♂.
O. Urvilliana, Boisd. Sp. Gén. Lép. p. 175.
Hab. New Ireland (*Paris Museum*).

b. *Pompeus* group.

6. ORNITHOPTERA REMUS, Cramer.

Papilio Remus, Cr. Pap. Ex. t. 135. f. A, t. 136. f. A (♀), t. 366. f. A, B (♂); Fab. Syst. Ent. iii. 1. p. 11.
O. Remus, Bd. Sp. Gén. Lép. p. 176. *Papilio Panthous* ♂, Clerck, Icon. t. 18 (♀).
Hab. Amboyna, Ceram, Gilolo, Morty Island, Sulla Island, Celebes (*Wall.*).

The specimens above quoted agree well with Cramer's figures. The female from the Sulla Islands differs only in having more yellow towards the anal angle of the lower wings. The specimens figured by Cramer in pls. 10, 11, under the name of "*Hypolitus*" seem to be a remarkable variety, in which the female has much of the character of the male. Messrs. Doubleday and G. R. Gray have adopted *Panthous* as the specific name of this insect; but this name was first used by Linnæus for the female of *Priamus* only, in the 10th ed. of the 'Systema Naturæ' (1758). Clerck (in 1759) adopted the name, but supposed he had found the male in the female of *Remus*. Linnæus, in Mus. Lud. Ulric. (1764), and in the 12th ed. of the 'Systema Naturæ' (1766), adopts this error, so far as referring to Clerck's two figures; but in both these works his description refers only to the female of *P. Priamus*, indicating that the supposed other sex (*P. Remus*) was not known to him personally. The name of *Panthous* must therefore altogether drop, it having been applied to this species only through a double error—first, that of Linnæus, in supposing his *Panthous* to be distinct from *Priamus*, and then that of Clerck, in thinking that a female *Remus* was the male of the Linnean *Panthous*.

7. ORNITHOPTERA HELENA, Linnæus.

♂. *P. Helena*, Cram. Pap. Ex. t. 140. f. A, B. *O. Helena*, Boisd. Sp. Gén. Lép. p. 177.
♀. *P. Amphimedon*, Cram. Pap. Ex. t. 194. f. A. *O. Amphimedon*, Boisd. Sp. Gén. p. 176.
Hab. Amboyna and Ceram (*Wall.*).

The females from these localities are always sooty, with the spots and markings on the hinder wings of a dull buff-colour even in the freshest specimens.

a. Local form *Bouruensis.*—*Male* exactly resembles the Amboyna specimens, except that the yellow patch is more variable in form and extent.

Female: nearly black, and with the markings on the lower wings almost as pure and deep yellow as in the males: size a little smaller than in the type.

Hab. Bouru (*Wall.*).

b. Local form *Papuensis.*—*Female:* sooty black, the two first branches of the subcostal nervure margined with whitish near their origin; markings of the lower wings of the same tint of orange-yellow as is *O. Helena* ♂, but not so glossy.

Male not known.

Hab. New Guinea, Salwatty (*Wall.*).

c. Local form *Celebensis*.—*Male*: wings a little more pointed than in *O. Helena*; yellow patch of lower wings extending nearer to the posterior margin, and bounded towards the abdominal margin by the first branch of the median nervure. Beneath, having the nervures between the discoidal cell and the outer border ashy-margined.

Female not known.

Hab. Macassar (Celebes) (*Wall.*).

Remarks.—Of these three local modifications of *O. Helena*, the first is very distinct in the female, but not separable in the male sex. Of the second and third, only one sex is known; and they may very probably prove to be well-marked species when more materials are obtained.

8. ORNITHOPTERA LEDA, n. s.

Male: upper wings elongate, triangular, glossy black, quite uniform and immaculate; the outer margin delicately white-marked at the termination of the nervures. Lower wings yellow, as in the allied species, with a black border about the same width as in *O. Pompeus* on the outer and abdominal margins, narrower on the inner margin; the posterior scalloping of the yellow patch not so deep as in *O. Pompeus*, and having a spot at the anal angle connected more or less with the margin.

The under surface differs from that of *O. Pompeus* by the ashy margins of the veins of the upper wings being entirely absent, and in having much less white on the outer edge. There are no submarginal spots except the anal one, much red at the base of the wings, and no black spots on the abdomen.

Female: this sex varies very much, some having the upper wings immaculate, while others have the veins about the end of the discoidal cell broadly margined with whitish. The marginal series of spots on the lower wings vary as they do in *O. Pompeus* and *O. Amphrisius*. The best distinction from *O. Pompeus* (♀) seems to be the more elongated wings, the less crenellated margin, and the more produced outer angle of the lower wings. The yellow patch is also of a deeper colour both on the upper and under surfaces.

Hab. Celebes (Macassar and Menado) (*Wall.*)

9. ORNITHOPTERA POMPEUS, Cramer.

P. Pompeus, Cr. Pap. Ex. t. 25. f. A (♂). *P. Minos*, Cr. Pap. Ex. t. 195, f. A (♀). *P. Heliacon*, Fab. Ent. Syst. 3. i. p. 19, 60.
O. Heliacon, Boisd. Sp. Gén. Lép. p. 178.

Hab. Sumatra, Borneo, Java, Lombock (*Wall.*), India (var.).

Remark.—The form that occurs in India, in its more elongate wings and darker colouring, approaches very closely to *O. Rhadamanthus*.

10. ORNITHOPTERA NEPHEREUS, G. R. Gray.

P. Astenous, Eschscholtz, Voy. Kotzebue, t. 4. f. A, B, C. (nec Fab.).
O. Nephereus, G. R. G., List of Lep. B. M. Papilionidæ, p. 6.
Hab. Philippine Islands.

Remark.—This is quite distinct from *O. Rhadamanthus*, Bd., with which it has generally been identified.

11. ORNITHOPTERA MAGELLANUS, Felder.

O. Magellanus, Feld. Lep. Nov. Phil. p. 11.
Hab. North of Luzon (Philippines).

Remark.—This fine species has a beautiful opalescent glow on the lower wings when viewed obliquely.

12. ORNITHOPTERA CRITON, Felder.

O. Criton, Feld. Lep. Fragm. p. 19.
Hab. Batchian, Ternate, Gilolo, Morty Island (*Wall.*).

13. ORNITHOPTERA PLATO, n. s.

Male: resembles *O. Criton* in the form and extent of the yellow patch, but the upper wings differ in having the outer half of a lighter tint; on the under surface this outer half of the wing is of a light ash-colour. Abdomen almost wholly black beneath. No red patches at the base of the wings, or any red collar.
Female unknown.
Hab. Timor (*Wall.*).

This is a very distinct species, though at first sight resembling several others. I obtained a single male specimen only.

14. ORNITHOPTERA HALIPHRON, Boisduval.

O. Haliphron, Bd. Sp. Gén. Lép. p. 181 (♂); Felder, Lep. Fragm. p. 37, Taf. ii. f. 2, 3 (♂, ♀).
Hab. Macassar (Celebes) (*Wall.*).

15. ORNITHOPTERA AMPHRISIUS, Cramer.

P. Amphrisius, Cr. Pap. Ex. t. 219. f. A; Godart, Enc. Méth. ix. p. 27, pt.
O. Amphrisius, Bd. Sp. Gén. Lép. p. 178.
Hab. Malacca, Java, Borneo (*Wall.*).

This may be readily distinguished from the allied species by the upper wings in the male being yellow-marked, and by the absence of red spots at the base of the wings beneath in both sexes.

c. *Brookeana* group.

16. ORNITHOPTERA BROOKEANA, Wallace.

O. Brookeana, Wall. Proc. Ent. Soc. 1855, p. 104; Hewitson, Ex. Butt. Papilionidæ, i. f. 1. *Papilio Trogon*, V. Voll. Tijdschrift voor Ent. 1860, p. 69, pl. 6.
Hab. Borneo (Sarawak) (*Wall.*), Sumatra (*Leyden Museum*).

Remarks.—I have been in much doubt about the position of this remarkable species, and was for some time inclined to place it among the Papilios. It agrees, however, far better with *Ornithoptera* in the form and stoutness of the wings, the long stout and curved antennae, the red collar and patches at the base of the wings beneath, the abdominal fold, and the flight and general appearance. It is powerful on the wing, and occasionally settles on the ground in damp sunny places. It inhabits the interior of North-west Borneo and the mountains of West Sumatra. The female is unknown. It is peculiar in the great length of the discoidal cell of the wings and its altogether unique style of coloration, and must be considered as the type of a distinct group of the genus *Ornithoptera*.

PAPILIO.

This is without doubt the finest and most remarkable genus of Diurnal Lepidoptera. About 360 species are now known, all, except ten, being tropical or subtropical. I have given at p. 23 the characters of the sections and groups into which I divide the Malayan species.

SECTION A.

a. *Nox* group.

17. PAPILIO NOX, Swainson.

P. Nox, Sw. Zool. Ill. pl. 102; Horsf. Lep. Ins. E. I. C. pl. 1. f. 1; Boisd. Sp. Gén. Lép. p. 277.
P. Neesius, Zink. Nov. Act. Acad. Nat. Cur. xv. t. 11. f. 1.
Hab. Java (♂, ♀) (*Wall.*), Penang (♂) (*Brit. Mus.*).

18. PAPILIO NOCTIS, Hewitson. Tab. V. fig. 1 (♂)*.

P. Noctis, Hewits. Proc. Zool. Soc. 1859, p. 423, pl. 66. f. 5 (♂).

Male: differs from the same sex of *P. Nox* by the broader apex of the fore wings, and by the hind wings being more elongate, more glossy, and especially by the entire non-dentated hinder margin.

Hab. Borneo (Sarawak) (*Wall.*), (♂, ♀ Mus. nost.)

19. PAPILIO EREBUS, Wallace.

P. Nox, var., De Haan, Verh. Nat. Gesch. t. 5. f. 3 (♀).
Hab. Malacca (*Wall.*), Banjermassing, Borneo (*De Haan*).

Remarks.—I am somewhat doubtful of the species, the female only being known; but it differs so strikingly from the same sex of *P. Nox* and *P. Noctis* (the former of which seems very constant), that I think it better to separate it in order to draw attention to other specimens that may exist in collections. It differs from *P. Nox* (♀) by its narrower and more elongate hind wings, which are black, glossed with steel-blue; the fore wings are black, with the veins beyond the cell clearly white-margined. The lower margin is also much less strongly dentated.

* In all the Plates, the wings on one side of each figure are detached from the body, and represent the *under surface* of the same insect. In one case only (Tab. VII. f.1.) the upper surfaces of two varieties of the same species are given.

20. PAPILIO VARUNA, White.

♀. *P. Varuna*, Wh., Entomologist, 1842, p. 280; Westw. Ann. Nat. Hist. ix. p. 37. *P. Chara*, Westw. Arc. Ent. pl. 66. f. 2.
♂. *P. Astorion*, Westw. Ann. Nat. Hist. ix. p. 37; Arc. Ent. pl. 66. f. 1.
Hab. Pulo Penang, **Sylhet**.

21. PAPILIO SEMPERI, Felder.

P. Semperi, Feld. Lep. Nov. Philipp. pp. 1, 11.
Hab. Luzon, **Philippines** (♂, ♀).

N.B. The *Philoxenus* group peculiar to India follows on after these.

b. *Coon* group.

22. PAPILIO NEPTUNUS, Guérin.

P. Neptunus, Guér. Deless. Voy. dans l'Inde, p. 69, t. 19 (*P. Saturnus*).
Hab. Malacca, Borneo (♂, ♀) (*Wall.*).

23. PAPILIO COON, Fabricius.

P. Coon, Fab. Ent. Syst. iii. 1. pp. 10, 27; Don. Ins. China, pl. 24. f. 1; Lucas, Lep. Ex. t. 6. f. 2; Boisd. Sp. Gén. Lép. p. 201.
Hab. Java, Sumatra (*Wall.*), Borneo (*De Haan*).

Remarks.—The specimens from Sumatra are constantly larger than those from Java. The Indian form, in which the markings are red instead of yellow, with other differences, I consider a distinct species, for which I propose the name of *P. Doubledayi*, after the late Mr. Edward Doubleday of the British Museum [*].

c. *Polydorus* group.

24. PAPILIO POLYDORUS, Linnæus.

P. Polydorus, L.; Clerck, Icon. t. 33. f. 3. *P. Leobates*, Reinw. Verh. Nat. Gesch. Zool. t. 6. f. 3 (♀).
Hab. Ceram, Matabello Island, Bouru, Batchian (♂, ♀) (*Wall.*).

Local form or variety *a.*—The white markings on the fore wings forming a patch below the cell; red spots on the hind wings nearer to the posterior margin and that next the anal angle larger.

Hab. Ke Island, Aru Island (♂, ♀) (*Wall.*).

---

[*] PAPILIO DOUBLEDAYI, Wallace. (*P. Coon*, var., B. M. Cat.)

Above: upper wings as in *P. Coon*, but the base darker. Lower wings broader than in *P. Coon*; the white spot in the cell toothed below, and divided by one or two faint blackish lines, cut off at the middle of the cell by the black triangular basal patch. The marginal spot next within the tail wanting; the two anal spots, end of abdomen, and its rings (which are yellow in *P. Coon*) red; collar behind the eyes and palpi (which are black in *P. Coon*) also red.

Beneath: base of lower wings broadly black; white spots all much broader and rounder than in *P. Coon*; sides of the thorax, end of the abdomen, and the marginal spots in the caudal and anal region red.

The female differs in a corresponding manner from *P. Coon* ♀. Size about the same.

Hab. Moulmein, Assam.

25. PAPILIO LEODAMAS, n. s. Tab. V. fig. 2 (♂).

P. *Polydorus*, in Brit. Mus. List of Papilionidæ, p. 10.

Male. Above, glossy black, upper wings immaculate (the veins pale-margined in the female). Lower wings with a rounded white spot divided into six parts by fine nervures, of which the outermost and that in the cell are sometimes reduced to points; marginal row of red spots obscured with black, and but faintly indicated.

Beneath, the white patch has a small red spot attached to the part next the anal angle; and the marginal row of six red spots are clearly marked, that at the anal angle being twice the size of the rest. Wings short, much rounded, scarcely or not at all produced in the caudal region.

Expanse of wings $3\frac{3}{4}$ in. to 4 in.

Hab. New Guinea, Mysol (♂, ♀) (*Wall.*), Rockingham Bay (Australia), (*Brit. Mus.*, ♀).

26. PAPILIO DIPHILUS, Esper.

P. *Diphilus*, Esp. Ausl. Schmett. t. 40. f. 1. P. *Polydorus*, Boisd. Sp. Gén. Lép. p. 267; and most authors.

Hab. Java, Malacca (*Wall.*), Philippine Islands, India.

Remarks.—The specimens from Manilla are larger, and the females paler-coloured, than those from other localities, all of which have slight characteristic peculiarities; but they also vary in the individuals from each locality, so that no perfect segregation of local forms has yet taken place.

27. PAPILIO ANTIPHUS, Fabricius.

P. *Antiphus*, Fab. Syst. Ent. iii. 1. pp. 10-28; Boisd. Sp. Gén. Lép. p. 266.

Hab. Sumatra, Borneo, Lombock, Java (*Wall.*), Philippine Islands.

Remarks.—The Philippine form (P. *Kotzebuea*, Eschsch.) is rather larger and of a more uniform glossy black than those from other localities. P. *Theseus*, Cram., has been erroneously supposed to be the female of this species, whereas it is the female of one of the *Pammon* group, belonging to a different section of the genus. De Haan figures P. *Theseus* as P. *Antiphus* ♀, in Verh. Nat. Gesch. t. 8. f. 2. As has been already pointed out, P. *Theseus* mimics this species.

28. PAPILIO POLYPHONTES, Boisduval.

P. *Polyphontes*, Bd. Sp. Gén. Lép. p. 268. P. *Heyeman*, G. R. G., List of Papilionidæ in B. Mus.

Hab. Celebes, Batchian, Morty Isl. (♂, ♀) (*Wall.*).

Remarks.—The markings vary from pure white to a smoky tint; but otherwise all the specimens from the above localities agree. De Haan gives (Verh. Nat. Gesch. t. 8. f. 4) a female of one of the *Pammon* group as P. *Polyphontes* ♀.

29. PAPILIO ANNÆ, Felder.

P. *Annæ*, Feld. Lep. Nov. Philipp. p. 1.

Hab. Mindoro (Philippines).

30. PAPILIO LIRIS, Godart.

P. Liris, God. Enc. Méth. iv. p. 72; Boisd. Sp. Gén. Lép. p. 268; De Haan, Verh. Nat. Gesch. p. 38, t. 1. f. 3 (♀).
Hab. Timor (*Wall.*), N.W. Australia (*Brit. Mus.*).

Remarks.—The Australian specimens are smaller. The female of *P. Œnomaus* mimics this species, as has been already mentioned (p. 22). Both species were taken by myself on the same spot, and, though such large and conspicuous insects, they could never be distinguished without a close examination after capture. The female of this species differs very little from the male, being rather larger, with broader wings and less vivid coloration.

SECTION B.

d. *Ulysses* group.

31. PAPILIO ULYSSES, Linnæus.

P. Ulysses, L., Cramer, Pap. Ex. t. 121. f. A, B (♂), t. 122 A (♀). *P. Diomedes*, Boisd. Sp. Gén. Lép. p. 202.
Hab. Amboyna, Ceram (♂, ♀) (*Wall.*).

Remark.—The largest specimens of this glorious insect are found in the island of Amboyna, where it is rather common, hovering about the forest pathways. It sometimes visits the gardens in the town of Amboyna.

32. PAPILIO PENELOPE, n. s.

Male: rather smaller than *P. Ulysses*. Upper wings with six black cottony patches, and all separate from each other; whereas in *P. Ulysses* there are seven, and the four lower ones are always united at their margins. The blue colour fills the discoidal cell, and generally extends beyond it at the extremity; the upper disco-cellular nervure not black-bordered as in *P. Ulysses*. Lower wings with the blue colour extending further along the abdominal margin, and not quite so far towards the outer angle.

Female: has the blue colour of the same form and extent as in *P. Ulysses* ♀, but of the same bright tint as in the male; the marginal lunules more deeply curved.

Expanse of wings 5 inches.

Hab. New Guinea, Waigiou, Aru Is. (♂, ♀) (*Wall.*).

Remark.—As all the other forms closely allied to *P. Ulysses* have received names (*Telemachus*, Montr., *Chaudoiri*, Feld., *Telegonus*, Feld., and *Ulyssinus*, Westw.), I have also given one to this form peculiar to New Guinea and the Papuan Islands, the distinctive characters of which, though very slight, seem sufficiently constant.

33. PAPILIO TELEGONUS, Felder.

P. Telegonus, Feld. Lép. Fragm. p. 50.
Hab. Batchian, Gilolo (♂, ♀) (*Wall.*).

Remark.—A very distinct species, separated from *P. Ulysses* by the extent of the cottony patch on the upper wings, and by the different form and colour of the blue markings.

34. PAPILIO TELEMACHUS, Montrouzier.

P. Telemachus, Mont. Ann. de la Soc. d'Agriculture de Lyon, 1856, p. 395.
Hab. Woodlark Isl. (S. E. of New Guinea).

Remark.—This is a small species (exp. 4 in.), with less blue on the lower wings.

c. *Peranthus* group.

35. PAPILIO PERANTHUS, Fabricius.

P. Peranthus, Fab. Syst. Ent. iii. 1. p. 15; Don. Ins. China, pl. 26; Lucas, Lep. Ex. t. 12. f. 2; Boisd. Sp. Gén. Lép. p. 205.
Hab. Java, Lombock (*Wall.*).

36. PAPILIO PERICLES, n. sp. Tab. VI. fig. 1 (♂).

Wings more elongate, and upper wings more pointed, than in *P. Peranthus*.

Above black, the basal half of a silvery blue, greenish towards the base of the costa, and purplish on the outer margin, where on the lower wings it shades off into separate scales. On the submedian and two lower branches of the median nervure are elongate black cottony patches as in *P. Ulysses*, the lower ones joined at the base, the upper one separate; above these the outer margin is of a brown black, with a few atoms of yellow and blue scales towards the apex; the blue colour extends beyond the discoidal cell of the upper wings in a line parallel with the outer margin, on the lower wings it rounds away to the anal angle, and below it are five submarginal lunules of blue atoms, the outer one almost obsolete, and that next the tail largest and most deeply coloured. Thorax and body green.

Beneath as in *P. Peranthus*, but the posterior range of lunules margined with brilliant blue and orange brown.

Expanse of wings 3½ inches.

Hab. Timor (♂) (*Wall.*).

37. PAPILIO PHILIPPUS, Wallace. Tab. VI. fig. 3.

P. Peranthus, var. A. Boisd. Sp. Gén. Lép. p. 204.

Above: basal half of the wings of a rich green-blue, the rest black, with a triangular patch at the apex of the uppers, formed of green atoms situated between the nervures; on the lower wings six large submarginal lunules, the lowest of which sends out some green atoms along the tail. The black cottony spot is of a different form from that of *P. Peranthus*, the separate patches being only joined in the middle, and two of them extending along the nervures in a point nearly to the discoidal cell.

Beneath brilliantly marked with lunules of buff, black, and blue.

Expanse of wings 4½–5 inches.

Hab. Moluccas (*Wall.*).

Remarks.—My specimen from Ceram is of a greener tinge, and the colour extends a little beyond the end of the discoidal cell; that from Batchian is smaller, of a bluer tinge, and the colour of less extent. The species seems to be very rare.

38. PAPILIO MACEDON, Wallace.   Tab. VI. fig. 2 (♂).

P. Peranthus, var. B., Boisd. Sp. Gén. Lép. p. 204.

Boisduval's description sufficiently shows the remarkable differences of form, size, and colouring which this species presents, compared with that of which he considers it a variety. The female agrees with the male, except that the colours are a little less brilliant, and the cottony patches of the fore wings are absent.

Expanse of wings, ♂, 5 inches; ♀, 5-6 inches.

Hab. Macassar, Menado (Celebes) (*Wall.*).

39. PAPILIO BRAMA, Guérin.

P. Brama, Guér. Rev. Zool. 1840, p. 43, t. 1. f. 3, 4.   *P. Palinurus*, De Haan, Verh. Nat. Gesch. pp. 5, 29.

Hab. Malacca, Sumatra (*Wall.*).

40. PAPILIO DÆDALUS, Felder.

P. Dædalus, Feld. Lep. Nov. Philipp. p. 2.

Hab. Luzon (Philippine Islands).

41. PAPILIO BLUMEI, Boisduval.   Tab. VI. fig. 4 (♂).

P. Blumei, Boisd. Sp. Gén. Lép. p. 206.

Hab. Menado (Celebes) (*Wall.*).   " Amboyna," *Bd.*, error of locality.

Remark.—This very fine species comes nearest to the last, but is of much larger size, and is conspicuous by its brilliantly coloured tails.

42. PAPILIO ARJUNA, Horsfield.

P. Arjuna, Horsf. Cat. Lep. E. I. Comp. pl. 1. f. 14; Boisd. Sp. Gén. Lép. p. 269.   *P. Arjuna*, var. *a.*, Brit. Mus. Cat. of Papilionidæ, p. 17.

Hab. Java, Borneo, Sumatra (*Wall.*).

The Bornean form differs from that of Java by its larger size, and on the under surface by the three middle lunules being formed of a violet line only, with scarcely a trace of red beneath it, and by the orange-red lunules both at the anal and outer angles being divided (not margined) by a violet line. The scales sprinkled at the base of the lower wings are white and blue, and are neither so dense nor do they extend so far as the yellowish scales of the Java specimens. In all these particulars the Sumatra specimens are somewhat intermediate, but approach most to those of Borneo. This is one of the examples which show the isolation of Java, notwithstanding its proximity to Sumatra.

f. *Memnon* group.

(N.B. The *Protenor* group of India is intermediate between this and the last group.)

43. PAPILIO MEMNON, Linnæus.   Tab. I. figs. 1 (♂), 2, 3, 4 (♀ s).

♂, *P. Memnon*, L., Cram. Pap. Ex. t. 91. f. C (♂); Boisd. Sp. Gén. Lép. p. 192.

♀, 1st dimorphic form, *P. Ancæus*, Cr. Pap. Ex. t. 222. f. A, B.

? *P. Laomedon*, Cr. Pap. Ex. t. 50. f. A, B.; De Haan, Verh. Nat. Gesch. p. 24, t. 3. f. 2.

♀, 2nd dimorphic form, *P. Achates*, Cr. Pap. Ex. t. 213. A.
Hab. Java, Sumatra (*Wall.*).

Local form *a*.—*Male*: border of posterior wings beneath narrow and of an ashy-blue colour.
Female: near *P. Anceus*, Cr., and *P. Laomedon*, Cr., but of an olive-ashy colour.
Hab. Borneo (*Wall.*).

Local form *b*.—*Male*: band on under side of posterior wings ashy; the spots large, with reddish-orange lunules between the two series, and below the four outer ones.
Hab. Lombock (*Wall.*).

Remarks.—The difference between the male and the 2nd form of female is so great, both in form and colouring, that they could not have been imagined to be the same, had they not been bred from the same larvæ. They have also been taken "*in copulâ*" by myself. Each form varies considerably, both individually and locally; yet there are none intermediate between the two. I consider them, therefore, as presenting a fine instance of dimorphism; and I also believe that the second form mimics *P. Coon*, for reasons which I have explained at p. 21.

44. PAPILIO ANDROGEUS, Cramer.

♂, *P. Androgeus*, Cr. Pap. Ex. t. 91. f. A, B.
♀, 1st dimorphic form, *P. Agenor*, L., Cr. Pap. Ex. t. 32. C. A, B.
♀, 2nd dimorphic form, *P. Achates*, Cr. Pap. Ex. t. 182. f. A, B ; *P. Alcanor*, Cr. Pap. Ex. t. 166, f. A.
Hab. Malacca (*Wall.*), India.

Remarks.—Ever since it was discovered that the insects figured by the old authors as *P. Anceus*, *P. Agenor*, *P. Achates*, &c. were varying females of *P. Memnon* and *P. Androgeus*, the whole of these were very naturally concluded to belong to one varying species. An examination of many extensive collections, however, has convinced me that the continental forms, on the one hand, and the insular ones, on the other, can be readily distinguished, and really form two very well-marked species. The red lunules at the anal region beneath characterize all specimens from India (*Androgeus*, Cr.), while these are entirely absent in all the insular specimens (*Memnon*, Cr.); and the same characteristic difference can be traced in a greater or less degree throughout all the infinitely varying female specimens. My specimen from Malacca has a faint trace only on the upper surface of the characteristic red mark at the base of the anterior wings; in other respects it resembles the continental individuals. This form mimics the Indian form of *P. Coon* (*P. Doubledayi*, Wall.).

45. PAPILIO LAMPSACUS, Boisduval.

P. Lampsacus, Bd. Sp. Gén. Lép. p. 190; De Haan, Verh. Nat. Gesch. p. 24, t. 2. f. 2.
Hab. Java (♂) (*Wall.*).

46. PAPILIO PRIAPUS, Boisduval.

P. Priapus, Bd. Sp. Gén. Lép. p. 190; De Haan, Verh. Nat. Gesch. p. 23, t. 2. f. 1.
Hab. Java (*Boisd.*), Sumatra (*Raffles*), Borneo (*De Haan*).

47. PAPILIO EMALTHION, Hübner.

♂, *Papilio Emalthion*, Hübn. Samml. Exot. ii. t. 117; *P. Emalthion*, Boisd. Sp. Gén. Lép. p. 196;
  P. Florides, Godt. Enc. Méthod. ix. p. 809; *P. Kirschoterus* in Eschsch. Voy. Kotzebue, t. 3. f. 5.
♀, 1st form, *P. Emalthion*, Cat. of Lep. Brit. Mus. pl. 5. f. 4.
♀, 2nd form, *P. Romanzovia*, Eschsch. Voy. Kotz. t. 2. f. 4; *P. Descombesi*, Boisd. Sp. Gén. p. 197;
  P. Florides, ♀, Godt. Enc. Méth. ii. p. 809.
Hab. Philippine Islands.

Remarks.—I have no doubt whatever that we have here another case of dimorphism, and I therefore unhesitatingly place these supposed species under one name. The male of *P. Emalthion* very closely resembles the next species (*P. Deiphontes*), and the 2nd form of female (*P. Romanzovia*, Eschsch.) as closely resembles the female of the same species; so that there can be no doubt that Godart was right in describing them as the sexes of his *P. Florides*. The female figured in the British Museum Catalogue is intermediate between these, but has more of the characters of the male; and it is to be remarked that both these forms of female have arrived in Europe accompanied by the same male. I am therefore obliged to reduce by one the hitherto received species of Philippine Papilios.

48. PAPILIO DEIPHONTES, n. s.

P. Deiphobus, var. A., Bd. Sp. Gén. Lép. p. 201.

♂. *Above*: exactly as in *P. Deiphobus*, but having a small tooth only in place of the tail, and the posterior band of a clear ashy blue.
Beneath: with the markings as in *P. Emalthion*, except that the red patch at the base of the upper wings is smaller.

♀. Also tailless, but resembling in markings the same sex of *P. Deiphobus*, the pale patch on the upper wings not extending into the discoidal cell.

Expanse of wings, ♂, 5½ inches; ♀, 5¾ inches.

Hab. Batchian, Gilolo, Ternate, Morty I-d. (*Wall.*).

49. PAPILIO DEIPHOBUS, Linnæus.

P. Deiphobus, L., Cramer, Pap. Ex. t. 181. f. A, B; Donovan, Ins. Ind. pl. 17. f. 2; Lucas, Lep. Ex. t. 11;
  Boisd. Sp. Gén. Lép. p. 200.
♀, *P. Alcandor*, Cr. Pap. Ex. t. 40. f. A, B.
Hab. Ceram, Amboyna, Bouru (*Wall.*).

Remark.—A simple variety of both this and the last species frequently occurs, in which all the markings on the under side are ochre-yellow instead of red.

50. PAPILIO ASCALAPHUS, Boisduval.

P. Ascalaphus, Bd. Sp. Gén. Lép. p. 209 (♂); De Haan, Verh. Nat. Gesch. p. 26. t. 4. f. 2 (♀).
Hab. Menado, Macassar (Celebes), Sulla Isl. (*Wall.*).

51. PAPILIO ŒNOMAUS, Godart.

P. Œnomaus, Godt. Encyc. Méth. ix. p. 72; Boisd. Sp. Gén. Lép. p. 199; De Haan, Verh. Nat. Gesch.
  p. 21, t. 4. f. 1 (♂), 2 (♀).
Hab. Timor (♂, ♀) (*Wall.*).

Remark.—As has been already noticed (p. 22), the female of this species closely resembles *P. Liris* ♀, in company with which it was captured.

g. *Heleaus* group.

52. PAPILIO SEVERUS, Cramer.

P. Severus, Cr. Pap. Ex. t. 227. f. A, B (♂), t. 278. f. A, B (♀); Boisd. Sp. Gén. Lép. p. 212.
Hab. Bouru, Ceram, Amboyna, Gilolo, Batchian, Aru Isl. (*Wall.*).

Remarks.—This species exhibits a large amount of simple variation, in the presence or absence of a pale patch on the uppers, in the brown submarginal marks on the lower wings, in the form and extent of the yellow band, and in the size of the specimens. The most extreme forms, as well as the intermediate ones, are often found in one locality and in company with each other, indicating that over the above range continual intermixture probably takes place, and thus prevents any one form from becoming specialized in a restricted area. The two following modifications of it, however, have acquired perfect stability, each in a large island situated on the extreme limits of the species. I therefore consider them to be distinct, though the actual differences are but small.

53. PAPILIO PERTINAX, n. s.   Tab. V. fig. 4 (♂).

Upper side: anterior wings rather more elongate and pointed than in *P. Severus*, dusky brown, with faint longitudinal rows of yellow scales in the cell, and with rather denser scales between the nervures beyond it; these are condensed into a narrow yellowish band parallel to the outer margin, and rather nearer to the cell than to it. Hind wings black, with three yellowish white subquadrate spots (the upper one smallest) situate between the outer angle and the discoidal nervule; beyond these and continued to the anal angle are a few very faint and minute groups of scales.

Under side as above, but the transverse band on the upper wings is whiter, and on the lower wings are seven submarginal brownish-yellow lunules, the middle ones least marked, and those at the outer and anal angles having above them a very small group of minute blue scales.

The female is paler-coloured, with the markings rather more diffused, and has on the under side an imperfect ocellus at the anal angle, a row of faint brown lunules extending to the three white spots, and two irregular lunules of blue atoms below those next the abdominal margin.

Expanse of wings, ♂, 4¼ inches; ♀, 5 inches.
Hab. Macassar (Celebes) (*Wall.*).

Remark.—This species was rather abundant near Macassar, in woody places, and was very constant in its markings and general aspect.

54. PAPILIO ALBINUS, n. s.   Tab. V. fig. 5 (♂).

Wings broader than in *P. Severus*, costa less arched, tail smaller, and the caudal margin less produced.

Upper side brown-black; anterior wings with very faint horizontal lines of yellowish scales in the cell; apical portion of the wing more thickly powdered between the nervures, the powdering fading away towards the outer angle. Posterior wings with a large yellowish-white patch, commencing close to the anterior margin, widening in the middle so as to cross the end of the cell, and ending in a triangle with prolonged apex at the abdominal margin; the outer edge of this spot is regularly angulated and scalloped; two very faint brown lunules occur next the anal angle; and the outer margin is rather broadly white-edged between the dentations.

Under side: the anterior wings have distinct greyish lines of scales between the nervures in the apical region; posterior wings not dotted with scales as in *P. Severus*, but with two or three single rows of scales in the cell only; the yellowish band consisting of a lunule next the upper margin, followed by three rhomboidal spots notched below, of which the middle one is the largest, then a roundish spot and a small horizontal mark; a row of seven submarginal lunules, of which the three middle ones are smallest and nearly obsolete, and that at the anal angle much the largest and, with the whitish marginal spot below it, forming an incomplete ocellus.

Expanse of wings $3\frac{1}{2}$–$3\frac{3}{4}$ inches.

Hab. New Guinea (δ) (*Wall.*).

55. PAPILIO PHESTUS, Guérin.

P. Phestus, Guér. Voyage de la Coquille, t. 14. f. 2; Bd. Voy. de l'Astrolabe, i. p. 41; Sp. Gén. Lép. p. 212.

Hab. New Guinea (*Paris Museum*).

56. PAPILIO HELENUS, Linnæus.

P. Helenus, L.; Cramer, Pap. Ex. t. 153. f. A, B; Lucas, Lep. Ex. t. 15. f. 2; Boisd. Sp. Gén. Lép. p. 211.
Hab. China ("*type*," *Cramer's figure*).

Local form *a*. Has more falcate wings and longer tail; the red marks at the anal angle beneath are divided by a violet-white mark.
Hab. North India.

Local form *b*. Same form of wings as the last, but smaller; the third and fourth lunules from the anal angle beneath very small or quite absent.
Hab. Java, Sumatra (*Wall.*).

57. PAPILIO HECUBA, n. s. Tab. V. fig. 3 (δ).

Upper wings falcate, and their outer margin much hollowed out, as in many of the Celebes butterflies.

δ. *Upper side*: the outer half of the anterior wings of a fine cottony texture, as in *P. Helenus*, but more marked; the red lunule at the anal angle wanting; the rest as in *P. Helenus*.

Under side: the lunules and ocelli are ochre-yellow instead of deep red, the two outer ones very small, the third almost obsolete, and the next two absent; the anal ocellus is bordered with blue above, and adjoining it is a blue lunule in the place of the red one in *P. Helenus*.

♀. *Upper side* of a browner colour; two orange-brown ocelli at the anal angle.
Under side: the lunules and ocelli all larger; the two intermediate ones entirely absent, as in the male.
Expanse of wings 5¼-5¾ inches.
Hab. Macassar, Menado (Celebes) (*Wall.*).

58. PAPILIO ISWARA, White.

P. Iswara, White, Entom. 1842, p. 280; Doub. and Hew. Gen. of Diurn. Lep. pl. 2. f. 1 (♀).
Hab. Penang, Malacca, Singapore, Borneo (♂, ♀) (*Wall.*).

59. PAPILIO HYSTASPES, Felder.

P. Hystaspes, Feld. Lep. Nov. Philipp. p. 12.
Hab. Luzon (Philippines).
This is the Philippine form of *P. Heleaus*.

60. PAPILIO ARASPES, Felder.

P. Araspes, Feld. Ent. Fragm. p. 17.
Hab. Philippine Islands.
This comes near to *P. Iswara*.

61. PAPILIO NEPHELUS, Boisduval.

P. Nephelus, Bd. Sp. Gén. Lép. p. 210; De Haan, Verh. Nat. Gesch. p. 29, t. 4. f. 4, ♂.
Hab. Malacca, Sumatra, Borneo (♂, ♀) (*Wall.*), Assam (*Brit. Mus.*).

h. *Pammon* group.

62. PAPILIO PAMMON, Linnæus. Tab. II. figs. 1 (♂), 3, 5, 6 (♀♀).

♂, *P. Pammon,* L.; Cram. Pap. Ex. t. 141. f. B; Boisd. Sp. Gén. Lép. p. 272.
♀, *P. Polytes,* L.; Cram. Pap. Ex. t. 265. f. A, B, C.
Hab. Malacca, Singapore (*Wall.*), China, India, Ceylon.

The continental specimens of *P. Pammon* have all considerably developed tails in both sexes; the insular specimens on the other hand, (which I treat as a separate species), have only a prominent tooth or very short tail in the males. The females also differ considerably, presenting an analogous but distinct series of forms. In the true *P. Pammon* the males are very constant; but the females exist under three distinct forms, each of them presenting more or less numerous varieties, viz.:—

1*st form* of female. Tab. II. fig. 3.
This exactly resembles the male, except in the possession of a distinct ocellus at the anal angle on the upper surface. Rarely a variety occurs having in addition a submarginal row of red lunules, indicating a slight approximation towards some varieties of the second form.

2*nd form* of female (*P. Polytes*). Tab. II. fig. 5.
This is by far the most common form of female. A variety of this rarely occurs, which

wants the red patch at the anal angle, and has the white patch formed of a row of spots all situated a little below the discoidal cell. This is the nearest approach to the first form.

3rd form of female (*P. Romulus*, Cram. Pap. Ex. t. 43. f. A; *P. Mutius*, Fab., Bois. Sp. Gén. p. 270; *P. Hector* ♀, De Haan). Tab. II. fig. 6.

This not uncommon Indian butterfly I consider to be a third form of the female of *P. Pammon*. I was first led to suspect this by finding that no males of it are known (the male and female from Ceylon, noted in the British Museum List, I have ascertained to be both females), nor have I been able to find any after an examination of the chief collections in England. It is also to be observed that it has been received from no locality which is not also inhabited by *P. Pammon*; there is no other known Indian butterfly that can possibly be the other sex of it; and lastly, it agrees very closely with the second form of female (*P. Polytes*) in all its details of form, texture, and neuration; and though at first sight having a very different aspect, specimens are to be found which by a very slight modification could be changed so as to resemble that form. I am therefore quite satisfied in my own mind that I am right in sinking this species into a form of *P. Pammon*. I have already stated my opinion that it mimics *P. Hector*, with which, however, it has no affinity. The resemblance was such as to induce De Haan to place it as the female of that species.

63. PAPILIO THESEUS, Cramer. Tab. II. figs. 2, 4, 7 (♀ ♀).

P. Theseus, Cr. Pap. Ex. t. 180. f. B (♀), Boisd. Sp. Gén. Lép. p. 276.
P. Antiphus ♀, De Haan, Verh. Nat. Gesch. p. 49, t. 8. f. 2; Brit. Mus. List. Pap. p. 12.
P. Polyphontes ♀, De Haan, Verh. Nat. Gesch. t. 8. f. 4.
P. Melanides, De Haan, Verh. Nat. Gesch. t. 8. f. 3 (♀).

Male like *P. Pammon* ♂, but smaller, and the tail always reduced to a projecting tooth.

Hab. Java, Sumatra, Borneo, Lombock, Timor (*Wall.*).

Local form *a.* Much larger; more falcate wings; a broad short tail.

Hab. Macassar (*Wall.*)

1st form of female. Tab. II. fig. 2.

Like the male, but with a very slightly marked blue and red ocellus at the anal angle. This is very rare in the islands. I found one specimen only in Timor, which I took "*in copulâ*" with a male almost exactly resembling it.

2nd form of female (*P. Polyphontes* ♀, De Haan). Tab. II. fig. 4.

Like the 2nd form of *P. Pammon* ♀; but has the pale portion of the anterior wing of a much lighter colour, and not extending so far towards the base of the wing; the white spot on the hind wings is more rounded, and has always a rather large portion within the cell. This form is to some extent local, not existing, I believe, in Sumatra, where it is replaced by the next.

Hab. Borneo, Java, Timor (*Wall.*).

3rd form of female (*P. Theseus*, Cr.; *P. Antiphus* ♀, De Haan). Tab. II. fig. 7.

This is well characterized by the entire absence of the white spot from the hind wings.

The red spots and lunules remain; but in some specimens only those in the anal region are visible, and these have a very close resemblance to *P. Antiphus*. This is also a local form, not occurring, I believe, in company with the last.

Hab. Sumatra, Lombock (*Wall.*).

4th form of female (*P. Melanides*, De Haan, Verh. Nat. Geseh. t. 8. f. 3).

I consider this to be an isolated modification of *P. Theseus*, Cr., peculiar to Borneo. It possesses all the characteristics of a female of this species.

Hab. Banjarmassing (Borneo) (*Leyden Museum*).

N.B. The 2nd, 3rd, and 4th forms of ♀ are all tailed, as in the females of *P. Pammon*.

64. PAPILIO ALPHENOR, Cramer.

P. Alphenor, Cr. Pap. Ex. t. 90. f. B (♀); Boisd. Sp. Gén. Lép. p. 274 (♂, ♀); *P. Ledebouria*, Eschsch. Voy. Kotz. t. 3. f. 7.

This is very closely allied to *P. Theseus*. The male is larger, has the caudal tooth scarcely perceptible, and on the under side has white instead of red marginal lunules. The female is tailed, much larger than *P. Theseus* ♀ form 2nd, from which it further differs by the white patch on the hind wings having the red markings blended with it, and more prominent.

Hab. Celebes, Bouru, Amboyna, Ceram (*Wall.*), Philippine Islands.

1st form of female (*P. Ledebouria*, Eschsch.).

Like the male, but with a brown tinge and an obscure anal lunule. This has been noticed only in the Philippine Islands.

2nd form of female (*P. Alphenor*, Cr.).

Distribution the same as the male.

3rd form of female (*P. Elyros*, G. R. Gray, B. M. List Pap. p. 26).

The white patch on the lower wings reduced to a small spot, or quite absent. There are many varieties of this, showing very instructively how such isolated forms of female as occur in the two preceding species may have been produced by simple variation followed by a " natural selection " of the forms best adapted to special conditions.

Hab. Philippine Islands (*B. M.*)

65. PAPILIO NICANOR, Felder, ' Voyage of the Novara,' pl. . . . f. *c*, *d*.

Male. Upper side:—like *P. Alphenor* ♂; but the band of white spots is broader and more regular, and there is a row of four white submarginal lunules.

Under side as in *P. Alphenor*; but the marginal spots of the upper wings, and the submarginal lunules of the lower wings, are larger and more distinct.

Female quite tailless, like the male. Upper side:—like *P. Alphenor* ♀; but the rufous anal spots are much smaller, not forming an ocellus at the anal angle, and they do not join the white central patch.

Under side, differs from *P. Alphenor* in nearly the same manner as on the upper side.

Hab. Batchian, Gilolo, Morty Island (*Wall.*).

Remarks.—The absence of tails in the female, and the white submarginal lunules in the male, distinguish this at a glance from all its allies. It has a comparatively restricted range, and is very constant in both sexes. The plate sent me by Dr. Felder is not numbered.

66. PAPILIO HIPPONOUS, Felder *.

P. Hipponous, Feld. Lép. Nov. Philipp. p. 12; *P. Dirouus*, B. M. List (no description).
Hab. Luzon, Mindanao (Philippines).

67. PAPILIO AMBRAX, Boisduval.

P. Ambrax, Bd. Sp. Gén. Lép. p. 218; Voy. au Pôle Sud, Lép. t. 1. f. 3, 4 (♂); De Haan, Verh. Nat. Gesch. t. 7. f. 2 (♀). *P. Orophanes*, Boisd. Sp. Gén. p. 275 (♀).
Hab. Mysol, Salwatty, Dorey (*Wall.*).

Remark.—I believe that two, if not three, well-marked forms or species have been mixed up under the name of *P. Ambrax*, as I have endeavoured to show by the references. My specimens of the two sexes of each show a uniformity of character in each locality.

68. PAPILIO AMBRACIA, Wallace.

P. Ambrax, Bd.; De Haan, Verh. Nat. Gesch. t. 7. f. 1 (♂).

Male. Differs from *P. Ambrax*, Bd., by the ashy-white patch at the apex of the anterior wings.

Female. Has a large, roundish, white patch on the anterior wings, extending from the discoidal cell to the hinder angle. The red lunules on the hind wings are smaller. Same size as *P. Ambrax*.

Hab. Waigiou (♂, ♀) (*Wall.*).

69. PAPILIO EPIRUS, n. s.

Male. Above:—anterior wings as in *P. Ambrax*; posterior wings more elongate, the white band much narrower, notched behind at the nervures, with the portions between regularly rounded; the part which crosses the cell is cut by black nervures, and there is an oblique red mark at the anal angle.

Beneath:—with a submarginal of seven lunules on the hinder wings, the one above the anal angle very large; whereas the last two species have one small lunule only beneath, at the anal angle.

Female. Is probably that figured in 'Voy. au Pôle Sud,' Lép. t. 1, f. 5, which resembles most the female of *P. Ambracia*, but differs in the form of the white and red patches. It is said to be from "the coasts of New Guinea"; but as the expedition touched at the Aru Islands, it is very probable that there is an error of locality, as I have ascertained to be very often the case in the indications furnished by these and other 'Voyages.'

Hab. Aru Islands (*Wall.*).

* Having obtained a specimen of this insect while these sheets are passing through the press, I find that it should have been placed next to *P. Severus*.

70. Papilio Dunali, Montrouzier.

P. Dunali, Mont. Ann. Soc. d'Agricult. de Lyon, 1856, p. 394.
Hab. Woodlark Island (S.E. of New Guinea).
Remark.—This seems closely allied to the last species.

i. *Erecetheus* group.

71. Papilio Ormenus, Guérin. Tab. III. figs. 2 (♂), 1, 3, 4 (♀ ♀).

P. Ormenus, Guér. Voy. de la Coquille, pl. 14. f. 3; Boisd. Sp. Gén. Lép. p. 244.
P. Erecetheus, var., Voy. au Pôle Sud, Lép. t. 1. f. 1, 2.
P. Amanga, Boisd. Sp. Gén. p. 216, ♀ (*P. Onesimus*, Hew. Ex. Butt. Pap. iii. f. 8).
Hab. Waigiou, Aru Isl., Ké Isl., Matabello and Goram Isl. (*Wall.*).

This belongs to a remarkable group of Papilios inhabiting the Austro-Malayan region, and which are especially interesting as exhibiting a good instance of polymorphism, the females being of two or three distinct forms.

The male in this species is characterized by the small amount of marking on the under surface.

1st *form* of female. Tab. III. fig. 1.

Almost exactly intermediate between the male and the normal female, which resembles *P. Erecetheus* ♀.

Upper side brown-black; a band of four whitish-yellow spots across the anterior wings beyond the cell, the upper one of the same size and position as in the male, the 2nd and 3rd elongated towards the cell, the 4th rather shorter than the 3rd, and immediately beneath it. Posterior wings with a central patch of a pale sulphur-yellow just crossing the end of the cell, and separated below into five truncate lobes; below this, and next the anal margin, are two irregular blue lunules, with a red lunule at the anal angle and a smaller one lower down beneath the second blue lunule.

Under side as above; on the hind wings the upper half of the yellow patch is dusky, and there is a complete submarginal series of seven red lunules.

Hab. Waigiou (a single specimen) (*Wall.*).

2nd *form* of female. Tab. III. fig. 3.

Resembles very closely *P. Erecetheus* ♀; but the white patch on the hind wings does not cover so much of the cell, and the two middle lobes are much elongated posteriorly, and separated by wedge-shaped spaces; the blue lunules are but slightly marked, and do not exceed two in number.

Under side:—differs from *P. Erecetheus* in the white patch never reaching the anterior margin of the hind wings. In a specimen from Waigiou, the four middle lunules are nearly white. This may be considered the typical form of female, as it occurs everywhere in company with the male.

3rd *form* of female (*Amanga*, Bd.). Tab. III. fig. 4.

I have three specimens of this form from three of the localities in which the male occurs. They differ slightly from each other, but agree generally with the figure and description above quoted. An allied form of female (of the next species) was observed closely followed by two males of the ordinary form; they were watched for some time, the males hovering over the females in the manner usual before pairing; and the three were then captured at one stroke of the net. This occurred three years after the capture of the specimen figured by Mr. Hewitson, and at once convinced me that these puzzling specimens were an additional form of female to a well-known male. The fact that the only females known of an allied species (*P. Tydeus*) are intermediate between these forms confirms this determination.

Hab. Aru Island, Mysol, Goram Isl. (*Wall.*)

72. PAPILIO PANDION, n. s.

Male. Closely resembles *P. Ormenus*, but presents the following differences:—

Upper side:—the band of spots across the fore wings is faintly marked, or more frequently quite absent; the grey lines bordering the nervures at the apex are more distinct; on the hind wings, the first three indentations of the whitish patch are followed by faint powdered lunules of the same colour.

Under side:—the apex of the fore wings is strongly marked with grey lines between the nervures, but has generally no spots; on the hind wings there is a curved submarginal band of lunules across the wing, viz., at the anal angle a large irregular red lunulate spot with a blue and a grey mark above it—2nd, a larger grey lunule with an angular blue mark below it, and a red lunule nearer the margin—3rd, a similar grey lunule and blue mark—4th, a larger grey lunule, and a smaller blue mark with a faint red lunule below—5th, a grey lunule and a faint blue dash below—6th, a blue lunule with a faint grey mark above—7th, a blue lunule with a very faint mark above it. These vary somewhat in different specimens, but the whole series can always be traced.

1st form of female.

Scarcely distinguishable from the typical female of the last species: the blue lunules on the under surface form a complete series, almost as in *P. Erecthens* ♀.

Hab. New Guinea, Salwatty, Mysol Island (with the male) (*Wall.*).

2nd form of female.

Upper surface:—fore wings as in *P. Onesimus*, Hew.; hind wings yellowish white, a broad black border along the anterior, and a narrow one along the posterior margin, two yellowish lunules near the outer angle, anal angle pale yellow, then an oblong black spot with a bluish mark in its upper part, followed by a second (half-obliterated) black spot.

Under surface with the same markings; but there are a series of six blue angulated marks upon a black ground, the two intermediate ones being smaller and less distinct. Abdomen yellow; under side black.

Hab. Dorey (New Guinea) (*Wall.*)

Remarks.—This specimen was taken in company with two males, as before mentioned. An insect, described by M. Montrouzier as the female of his *P. Godartii* (from Woodlark Island), agrees very closely with this, and is no doubt the female of the same species, or a closely allied one which he puts in his list as *P. Ormenus*. The fact, therefore, that this peculiar pale form of female *Papilio* has been found in five islands, from no one of which is a male insect known which can be mated with it, except those of the *Ormenus*-form (which always occur in the same places), may, in conjunction with the observation already given of the companionship of the two forms, be taken to prove that this is really a case of polymorphism. I believe also it will be found that these extreme departures from the typical form of a species are connected with mimetic resemblances and the safety of the individuals. We have already seen that the extreme forms of *P. Memnon* ♀ and *P. Pammon* ♀ respectively resemble other species which from their habits and abundance seem to have some peculiar immunity from danger. In this case also there is a resemblance to quite a different family of butterflies, the Morphidæ. In form, coloration, and general appearance these pale-coloured Papilios resemble species of the genus *Drusilla*; and the same genus is also imitated by other butterflies—one of these, *Melanitis Agondas* ♀, having been actually confounded with *Drusilla bioculata* as the same species, so great is the resemblance. This fact of species of several genera imitating the Drusillas would indicate that they have some special immunities which make it advantageous to other insects to be mistaken for them; and their habits confirm this opinion. They have all a very similar style of dress, and fly very slowly, low down in damp woods, often settling on the ground or on rotten wood; and they are exceedingly abundant in individuals. Now these are the general characteristics of all groups which are the subjects of imitation; and we may therefore presume, when we see forms departing widely from the general appearance of their close relations, and resembling closely other groups with which they have no affinity, that what we must call *accidental* variations have been accumulated and rendered definite by natural selection for the protection and benefit of those forms.

73. PAPILIO TYDEUS, Felder. Tab. IV. figs. 3 (♂), 2 (♀).

P. Tydeus, Feld. Lep. Fragm. p. 52 (♂).

Female.—Upper side dusky brown; fore wings with the central portion below the cell nearly white; hind wings with the basal two-thirds white, with an irregular and obtusely dentated margin, and edged with ochre-yellow; the rest black, with a submarginal row of seven broad yellowish lunules, and above those nearest the anal angle three irregular blue patches.

Under side nearly as above; the white space on the upper wings is more extensive and better defined; the marginal lunules are dilated so as to form a crenellated band, and the blue marks are increased to six or seven in number. Head and thorax dusky; abdomen yellowish.

Hab. Batchian, Morty Island (*Wall.*).

Remark.—The female, which seems to be of only one form in this species, is especially interesting as being allied to the pale-yellow form of *P. Ormenus* and *P. Pandion*.

74. PAPILIO ADRASTUS, n. s. Tab. IV. fig. 1 (♀).

Male.—Upper side, like *P. Ormenus* ♂; but has the band of the hind wings narrower, not crossing the cell, and more pointed towards the anal angle.

Under side with a single red anal spot, and three blue lunules beyond it.

Female.—Upper side brown black; anterior wings with the apical half browner, a whitish patch around the end of the cell, and an ovate spot within it; posterior wings with a small central whitish patch more or less tinged with ochreous; a submarginal row of very large deep-red lunules, that at the anal angle forming an irregular ocellus bordered above with pale blue, and a few blue atoms on the side of it. Indentations of all the wings broadly margined with ochreous.

Under side:—the white patch of the anterior wings larger and well defined, and continued by smaller and fainter patches to the outer angle; posterior wings with the small central patch and marginal lunules as above, with the addition of a faint row of angulated blue marks between them.

Wings elongated posteriorly, and somewhat angulated at the termination of the first median nervure.

Expanse of wings, ♂, 5¼ inches; ♀, 6 inches.

Hab. Banda Island (*Wall.*).

Remarks.—This species is near *P. Ormenus* in the male, but approaches *P. Gambrisius* in the female, which differs from all others in this group by its dark colouring and the short narrow band on the hind wings. A male and two females were obtained in the small island of Banda.

75. PAPILIO GAMBRISIUS, Cramer.

P. Gambrisius, Cr. Pap. Ex. t. 157. f. A, B (♂); Boisd. Sp. Gén. Lép. p. 213.
P. Drusius, Cr. Pap. Ex. t. 229. f. A, t. 230. f. A (♀); Boisd. Sp. Gén. Lép. p. 218.
Hab. Amboyna, Ceram, Bouru (*Wall.*).

Remarks.—The males of this fine species are not uncommon in Ceram, and in hot weather come down to the beach and settle on the wet sand. The females, however, are very rare; I obtained one in the mountainous forests of Ceram, and this is, I believe, the only fine and perfect specimen now in Europe.

Expanse of male 5½–6½ inches, of female 7 inches.

76. PAPILIO AMPHITRION, Cramer.

P. Amphitrion, Cr. Pap. Ex. t. 7. f. A, B; Boisd. Sp. Gén. Lép. p. 217.
Hab. Celebes?

Remarks.—The habitat of this rare species is doubtful. Cramer says, "America;" Godart, "Amboyna;" but I believe its true locality will be found to be Celebes. It forms a transition to the next species.

77. PAPILIO EUCHENOR, Guérin.

P. Euchenor, Guér. Voy. de la Coquille, t. 15. f. 3 (♂); *P. Axion,* Boisd. Sp. Gén. Lép. p. 46 (♂).

Female.—Similar to the male; but the markings are all of a dull ochre-yellow, and the second and third spots, reckoning from the inner margin of the upper wings, are almost entirely wanting. This sex is much rarer than the male.

Hab. New Guinea, Aru Island, Ké Island (*Wall.*).

78. PAPILIO GODARTII, Montrouzier.

P. *Godartii*, Montr. Ann. Soc. d'Agric. de Lyon, 1856, p. 394.
Hab. Woodlark Island.

Remark.—Closely allied to the last; perhaps a variation only.

k. *Demolion* group.

79. PAPILIO DEMOLION, Cramer.

P. *Demolion*, Cr. Pap. Ex. t. 89. f. A, B; P. *Cresphontes*, Fabr.; Boisd. Sp. Gén. Lép. p. 220.
Hab. Java, Borneo, Sumatra, Singapore (*Wall.*), Moulmein (*Brit. Mus.*).

80. PAPILIO GIGON, n. s. Tab. VII. fig. 6 (♀).

"*P. Gigon*," List of Papilionidæ in Brit. Mus. p. 27 (no description).

Much larger than *P. Demolion*; costal margin of the fore wings very much arched from the base; tail proportionally shorter.

Upper side:—markings as in *P. Demolion*, with the following differences. In the cell of the fore wings are four longitudinal curved greyish-yellow lines; the yellow band begins higher on the abdominal margin, and curves outward toward the tip, where the spots are obliquely elongate, and the three last distinctly notched; on the hind wings the lunulate spots are much deeper and are rather further from the margin, and the two spots at the outer angle (often obsolete in *P. Demolion*) are large and well marked.

Under side:—the markings resemble those of *P. Demolion*, but are stronger; the band of silvery spots is much more sinuate, and possesses an additional lunule above the outer angle; a patch of ochre-yellow covers the lower margin of the cell, extending a little along the nervures which radiate from it.

Abdomen blackish, with numerous stripes and spots of pale yellow.

Expanse of wings $4\frac{3}{4}$ to $5\frac{1}{4}$ inches.

Hab. Celebes, Sulla Island (*Wall.*).

Remark.—This was regarded by Boisduval as a large variety of *P. Demolion* (see Sp. Gén. Lép. p. 221); but it offers remarkable differences both in form and markings.

l. *Erithonius* group.

81. PAPILIO ERITHONIUS, Cramer.

P. *Erithonius*, Cr. Pap. Ex. t. 232. f. A, B.
P. *Epius*, Fabr.; Don. Ins. China, pl. 29. f. 2; Boisd. Sp. Gén. Lép. p. 238.
Hab. India, China (*type*).

Local form *a* (*Malayanus*).—The two spots on the lower margin of the cell of the hind wings wanting; anal spots redder, and the ocellus at the outer angle darker: two spots in cell of fore wings, as in the type; but in Flores specimens these approach so as almost to unite.

Hab. Singapore, Flores (*Wall.*), Manilla.

Local form *b* (*Sthenelus*, Macleay).—A single large spot in the cell of the fore wings; one small detached spot on the margin of the cell of the hind wings.

Hab. Goram Island (*Wall.*), Australia.

Section C.

m. *Paradoxa* group.

82. Papilio Paradoxa, Zinken.

Zelima Paradoxa, Zink. Beitr. Ins. Java, t. 15. f. 9, 10.
P. Paradoxa, Westw. Cab. Or. Ent. pl. 9. f. 1, 1*.
Hab. Java (*Wall.*).

Local form *a.*—*P. Paradoxa*, var., Hew. Proc. Zool. Soc. 1859, p. 422, pl. 67. f. 1 (♂), 2 (♀).

Hab. Borneo (*Wall.*).

Local form *b.*—Smaller; intermediate in the markings between the Java and Borneo forms; interior row of elongate marks on upper wings light blue, not descending to the outer angle.

Hab. Sumatra (*Wall.*).

Both sexes of this species closely resemble the corresponding sexes of *Euplœa Midamus*, Cr., which is very common in all the above-mentioned localities.

83. Papilio Ænigma, n. s. Tab. VII. fig. 3 (♂).

Size, form, and markings nearly the same as in *P. Paradoxa*.

Above:—purplish black, without any gloss or silky reflexions; a submarginal row of white spots on all the wings, more or less blue-edged on the upper wings, sometimes partially obsolete on the lower ones; one or two spots at the end of the cell, and a row of six or seven elongate marks beyond it, bright blue.

Beneath, the submarginal row of white spots only.

Female.—*P. Paradoxa*, var. A, Hewitson, Proc. Zool. Soc. 1859, p. 423, pl. 67. f. 3.

I put this as the female of the above with some hesitation, as it was not captured in the same island. It agrees, however, in the entire absence of gloss, and in the peculiar elongation of the outer angle of the lower wings.

Hab. Malacca, Sumatra (♂); Borneo (♀) (*Wall.*).

Female variety?—*P. Paradoxa*, var. B, Hewitson (Proc. Zool. Soc. pl. 66. f. 4), may be an extreme variation of this, but will more probably, when the male is discovered, prove to be a distinct species.

84. Papilio Caunus, Westwood.

P. Caunus, Wesw. Cab. Or. Ent. pl. 9. f. 2, 2*.
Hab. Sumatra, Borneo (♂, ♀) (*Wall.*), Java (*Leyden Mus.*).

Remarks.—My specimens have less white on the lower wings than is represented in Mr. Westwood's figure. The female is of a brownish colour, with the same white markings as the male, but without any blue tinge. This species is very like *Euplœa Rhadamoathus*, one of the most common butterflies in all the above-mentioned localities. It is undistinguishable from that insect on the wing, though it flies very slowly, like the species it mimics.

85. PAPILIO ASTINA, Westwood.

P. Astina, Westw. Cab. Or. Ent. pl. 9. f. 3.
Hab. Java (*Brit. Mus. ex Coll. Horsf.*).

86. PAPILIO HEWITSONII, Westwood.

P. Hewitsonii, Westw. Proc. Ent. Soc. 1864, p. 10.
P. Slateri ♀, Hew. Ex. Butt. Pap. pl. 4. f. 9; *P. Cauora*, B. M. List of Papilionidæ (no description).
Hab. Borneo (♂) (*Wall.*).

Remarks.—The last two species should probably form a distinct group, on account of the peculiar elongation of the cell of the lower wings. They both resemble dark species of *Euplœa*. *P. Slateri* is a quite distinct species from North India, to which Mr. Hewitson referred the present species as the female. All the specimens known of both species are, however, males.

n. *Dissimilis* group.

87. PAPILIO ECHIDNA, De Haan.

P. Echidna, De Haan, Verh. Nat. Gesch. p. 42, t. 8. f. 6; *Clytia dissimilis*, Sw. Zool. Ill. 2nd ser. pl. 120;
P. dissimilis, var., Brit. Mus. List of Papilionidæ.
Hab. Timor (♂, ♀) (*Wall.*).

Remarks.—This species has been confounded with *P. dissimilis*, from which it is very distinct, by the absence of the yellow marginal band beneath. It is also widely separated geographically from that species, which inhabits the continent of India only. The sexes are alike, as they are in *P. dissimilis*. *P. Panope*, L., which has been supposed to be its female, is a very distinct species, of which also both sexes exist in most collections.

88. PAPILIO PALEPHATES, Westwood.

P. Palephates, Westw. Arc. Ent. pl. 79. f. 1; *P. dissimilis*, var. b, Brit. Mus. List of Papilionidæ.
Hab. Philippine Islands.

SECTION D.

o. *Macareus* group.

89. PAPILIO VEIOVIS, Hewitson.

P. Veiovis, Hew. Ex. Butt. Pap. pl. 7. f. 26 (♂).
Hab. Menado (Celebes) ("*Coll. Hewitson.*").

Remark.—This fine new species has been recently received from Menado, and seems best placed in this group, near *P. Encelades*.

90. PAPILIO ENCELADES, Boisduval.

P. Encelades, Bd. Sp. Gén. Lép. p. 376; Hewitson, Ex. Butt. Pap. pl. 4. f. 10 (♂).
Hab. Macassar (Celebes) (*Wall.*).

91. PAPILIO DEUCALION, Boisduval.

P. Deucalion, Bd. Sp. Gén. Lép. p. 375; Hewitson, Ex. Butt. Pap. pl. 4. f. 11 (♀).
Hab. Macassar, Menado (Celebes) (*Wall.*).

Remarks.—At Macassar I took only males of *P. Encelades*, and females of *P. Deucalion* at the same spot (a half-dry river-bed), and therefore conjectured that they might be sexes of one species, although so unlike. Some years afterwards, however, I took at Menado a fine male of *P. Deucalion*, which only differs in its rather smaller size and brighter colouring.

92. PAPILIO IDÆOIDES, Hewitson.

P. Idæoides, Hew. Ex. Butt. Pap. pl. 1. f. 2.
Hab. Philippine Islands (♂) (*Brit. Mus.*).

Remark.—This singular species must closely resemble on the wing *Hestia Leuconoë*, from the same islands.

93. PAPILIO DELESSERTII, Guérin.

P. Delessertii, Guér.; Deless. Souvenirs, t. 17.
Hab. Pulo Penang (*Hope Museum, Oxford*).

Remark.—This resembles the species of *Hestia* and *Idæopsis*, from the same locality, and is intermediate in size. It has been confounded with the next.

94. PAPILIO DEHAANII, Wallace.

P. Laodocus, De Haan, Verh. Nat. Gesch. t. 8. f. 5 (nec Fab.); *P. Melanides*, Erichs. Archiv für Natur. 1843 (nec De Haan, 1839).
Hab. Malacca, Borneo (*Wall.*), Java (*Leyden Mus.*).

Remarks.—The Bornean specimens are rather larger, and have the yellow anal spot somewhat differently shaped. The two names which have been applied to this species having been preoccupied, I have named it after the first describer.

95. PAPILIO LEUCOTHOË, Westwood.

P. Leucothoë, Westw. Arc. Ent. pl. 79. f. 3; *P. Xenocles*, var., Brit. Mus. List of Pap.
Hab. Singapore, Malacca (*Wall.*). N. India.

96. PAPILIO MACAREUS, Godart.

P. Macareus, Godt. Enc. Méth. ix. pl. 76; Horsf. Desc. Cat. Lep. E. I. C. pl. 5. f. 1; Boisd. Sp. Gén. Lép. p. 374.
P. striatus, Zink. Beitr. Ins. Java, t. 14. f. 5.
Hab. Malacca (*Wall.*), Java (*Horsfield*), Borneo (*Leyden Mus.*).

This species closely resembles *Danais Aglaë*, Cr., found in the same islands.

97. PAPILIO STRATOCLES, Felder.

P. Stratocles, Feld. Lép. Nov. Philipp. p. 2.
Hab. Mindanao (Philippines).

98. PAPILIO THULE, n. s. Tab. VII. fig. 1 (♂).

Form of *P. Macareus,* but smaller.

Above:—brown-black, spotted and marked with greenish white; a row of spots near the outer margin of all the wings, and on the upper wings a second row between the first and the end of the cell, three or four others close to the cell, and 5–7 irregularly placed in the cell; the spot next the outer angle is double, and the two lower spots of the second row are continued indistinctly to the cell. The lower wings have a mark at the end of the cell, and five elongated spots radiating from it between the nervures.

Beneath:—brown, with the spots all whiter and more distinct. Neck with four white points; abdomen dusky, with pale lines on the sides and beneath.

Expanse of wings 3⅞ inches.
Hab. New Guinea (♂) (*Wall.*).

Variety or local form *a.*—Like the above, but with the discal spots of the lower wings united into a transverse band divided by fine nervures.
Hab. Waigiou Island (♂) (*Wall.*).

This species imitates *Danaïs sobrina,* Bd., a New Guinea species. The figure represents the upper surface of both forms of this insect.

p. *Antiphates* group.

99. PAPILIO ANTIPHATES, Cramer.

P. Antiphates, Cr. Pap. Ex. t. 72, f. A, B; Boisd. Sp. Gén. Lép. p. 248.
P. Pompilius, Fab.; Lucas, Lep. Ex. t. 22. f. 1; Godt. Enc. Méthod. ix. p. 49.
P. Alcibiades, Fab.; Godt. Enc. Méthod. ix. p. 49.
Hab. India, China ("*type*").

Local form *a.*—*Podalirius* Pompilius, Sw. Zool. Ill. 2nd ser. pl. 105.
Hab. Malacca, Sumatra, Java, Borneo (*Wall.*).

These differ from the type in the black apical portion not quite reaching the outer angle, and in the first and second bands on the upper wings not extending below the cell. The fourth band varies in extent, as does the amount of grey colouring in the caudal region.

100. PAPILIO EUPHRATES, Felder.

P. Euphrates, Feld. Lep. Nov. Philipp. p. 12; *P. Coretes,* Brit. Mus. List of Papilionidæ (no description).
Hab. Philippine Islands.

101. PAPILIO ANDROCLES, Boisduval. Tab. VII. fig. 5 (♂).

P. Androcles, Bd. Sp. Gén. Lép. p. 279.
Hab. Macassar (Celebes) (*Wall.*).

Remarks.—I only met with this magnificent species on one occasion, on the banks of a mountain-stream and on the sands close to a waterfall. When resting on the ground, the very long white tails are raised up at a considerable angle, and are very conspicuous.

102. PAPILIO DORCUS, De Haan.

P. Dorcus, De Haan, Verh. Nat. Gesch. Zool. t. 7. f. 4.
Hab. Gorontalo (N. Celebes) ("*Leyden Museum*").

103. PAPILIO RHESUS, Boisduval.

P. Rhesus, Bd. Sp. Gén. Lép. p. 253.
Hab. Macassar (Celebes) (*Wall.*). "Bengal," the locality given by Boisduval, is erroneous.

104. PAPILIO ARISTÆUS, Cramer.

P. Aristæus, Cr. Pap. Ex. t. 318, f. E, F; Boisd. Sp. Gén. Lép. p. 252.
Hab. Ceram, Batchian (*Wall.*).

105. PAPILIO PARMATUS, G. R. Gray.

P. Parmatus, G. R. Gray, Cat. Lep. Ins. Brit. Mus. pl. 3, f. 2.
Hab. Aru Islands, Waigiou (*Wall.*), Australia (*Brit. Mus.*).

Remarks.—The Aru specimen agrees almost exactly with the type specimen in the British Museum. The Waigiou insect is rather darker on the under surface, and has the black markings more sharply defined.

q. *Eurypylus* group.

106. PAPILIO CODRUS, Cramer.

P. Codrus, Cr. Pap. Ex. t. 179. f. A, B; Boisd. Sp. Gén. Lép. p. 228.
Hab. Amboyna and Ceram (*type*) (♂, ♀) (*Wall.*).

Local form *a* (*Gilolensis*).—Differs from the true *P. Codrus* in having always an additional semiovate spot below the submedian nervure, and in having a small round spot on the anterior margin of the lower wings beneath: it is also rather smaller.
Hab. Batchian and Gilolo (*Wall.*)

Subspecies *b* (*Celebensis*).—Fore wings in the male more attenuate, with the costal margin more curved than in true *P. Codrus*; upper surface more green and glossy; an additional large quadrate spot on the inner margin of the fore wings. Under surface lighter brown, the whitish marks near the anal angle wanting; a dark subtriangular band across the cell of the fore wings. Rather smaller than *P. Codrus*.
Hab. Celebes, Sulla Islands (*Wall.*).

Subspecies *c* (*Papuensis*).—Hind wings less elongate than in the true *P. Codrus*; macular band much broader, and reaching the inner margin of the upper wings, the lower portion divided by nervures only; the band continued on the lower wings by means of an obscure white fascia.

Beneath, the greenish white band continues on to the lower wings, but gradually fades away after reaching the cell. Expanse of wings 4¼ inches.

Hab. Waigiou, Aru Island (*Wall.*).

Remarks.—This approaches the next species. Subspecies *b* and *c* I consider to be really as distinct as many universally received species, differing in form and in several points of coloration. As, however, it is probable that there are forms in other islands which may present intermediate characters, I prefer retaining the whole under the old specific name.

107. PAPILIO MELANTHUS, Felder.

P. Melanthus, Feld. Lep. Nov. Philipp. p. 12.
Hab. Mindanao (Philippines).

108. PAPILIO EMPEDOCLES, Fabricius.

P. Empedocles, Fab. Ent. Syst. iii. 1. p. 70; Don. Ins. Ind. pl. 17. f. 1; Boisd. Sp. Gén. Lép. p. 229.
Hab. Borneo (*Wall.*).

109. PAPILIO PAYENI, Boisduval.

P. Payeni, Bd. Sp. Gén. Lép. p. 235; Van der Hoeven, Tijd. von Nat. Gesch. v. t. 8, f. 1, 2, 6.
Hab. Borneo (*Wall.*), Java (*Van der Hoeven*).

Remarks.—This remarkable species has been placed by Boisduval in a group by itself. It, however, agrees very closely in habits and structure with this group, and can hardly, I think, be separated, though very abnormal in colouring. *P. Evan*, Db., is a closely allied species from India; and *P. Gyas*, Westw., from the same country, is also nearly related, though it has been hitherto placed in another section of the genus.

110. PAPILIO SARPEDON, Linnæus.

P. Sarpedon, L.; Cram. Pap. Ex. t. 122. f. D, E.; Boisd. Sp. Gén. Lép. p. 235.
Chlorisses Sarpedon, Sw., Zool. Ill. 2nd ser. pl. 89.
Hab. Borneo, Sumatra (typical), New Guinea, Aru Is. (darker), Java (broader band) (*Wall.*).

Local form *a* (*Moluccensis*, Cram. Pap. Ex. t. 122. f. D, E).—Black, with the bands and spots rich blue.
Hab. Ceram, Batchian, Gilolo, Bouru (*Wall.*). (The Ceylon form closely resembles this.)

111. PAPILIO MILETUS, n. s. Tab. VII. fig. 2 (♂).

Wings larger and more falcate than in *P. Sarpedon*, costal margin abruptly curved near the base of the wing.

Above, black; macular band rich blue, very narrow, the spots on the upper wings all more or less rounded and separated by thick black bands; the marginal lunules large and angularly bent.

Beneath, the upper wings have a row of four pearly-white lunules from the outer angle; and there is one of the same colour at the outer angle of the lower wings, which have also an additional red spot on the margin of the cell, below the first branch of the subcostal nervure. Expanse of wings 4⅜ inches.

Hab. Macassar and Menado (Celebes) (*Wall.*).

Remarks.—I have separated this species from all the other forms of *P. Sarpedon*, because, while they differ in markings and colour only, this differs greatly in form as well as very strikingly in size, colour, and markings. I cannot conceive, therefore, why such a combination of distinctive peculiarities should not entitle it to specific rank.

112. PAPILIO WALLACEI, Hewitson.

P. Wallacei, Hew. Ex. Butt., " Papilio," iii. f. 7.
Hab. Aru Islands, Batchian (*Wall.*).

Remark.—This isolated species is very rare: I obtained a single male specimen in each of the above localities in the virgin forest.

113. PAPILIO BATHYCLES, Zinken.

P. Bathycles, Zink. Beitr. Ins. Java, p. 157, tab. 14. f. 6, 7; Boisd. Sp. Gén. Lép. p. 232.
Hab. Java, Borneo, Malacca (*Wall.*).

Remark.—The Indian form generally confounded with this I consider to be a very distinct species, for which I propose the name of *P. Chiron*, and add a description below *.

114. PAPILIO EURYPYLUS, Linnaeus.

P. Eurypylus, L.; Cram. Pap. Ex. t. 122. f. C, D; Boisd. Sp. Gén. Lép. p. 243.
Hab. Amboyna (type), Ceram, Bouru, Batchian, New Guinea (*Wall.*).

Remark.—The *male* has the abdomen above and abdominal margin white; the *female* blackish.

115. PAPILIO JASON, Esper.

P. Jason, Esp. Ausl. Schmett. t. 58. f. 5; *P. Jason*, L.? *P. Eurypylus*, var., Boisd. Sp. Gén. Lép. p. 233.
Hab. Malacca, Sumatra, Borneo (♂, ♀) (*Wall.*).

Remarks.—This species is readily distinguished from *P. Eurypylus* by the abdomen above, and the abdominal margin, being black in both sexes, by the smaller size, more pointed upper wings, and by the lower wings having a narrower band and larger spots

---

* PAPILIO CHIRON, n. s.
P. Bathycles (partly), Brit. Mus. List of Papilionidæ.

Very near *P. Bathycles*, Zinken. Fore wings rather broader at the tip; hind wings considerably less elongate posteriorly.

Above :—fore wings have the three larger green spots separated by broad black spaces, the first produced towards the base of the wing, the second notched above; the fourth spot in the cell much more linear. Hind wings have the green markings more elongate and narrower, and an additional narrow mark at the abdominal margin.

Beneath, the spots all separated by broad black lines; the abdominal stripe, which is quite wanting in *P. Bathycles*, larger than above; an ochre-yellow spot on the hind wings, near the base of the inner margin (absent in *P. Bathycles*); the submarginal pale spots larger, and the row of reddish-ochre spots less developed. Expanse of wings 3½ inches.

Hab. Assam, Sylhet.

of a deeper green colour. On the under surface the marginal lunules, the cell-spots, and sub-basal stripe are all larger.

Variety or dimorphic form *a.*—*Ecemon*, Bois. Sp. Gén. Lép. p. 234.

Hab. Malacca, Java, Sumatra, Borneo (♂, ♀). (*Wall.*)

This may be a distinct species, but is more probably a case of dimorphism. The two forms are absolutely identical, except that the red spot at the base of the lower wings beneath, in *P. Jason*, is constantly absent in *P. Ecemon*.

116. PAPILIO TELEPHUS, n. s. Tab. VII. fig. 4 (♂).

Larger than *P. Eurypylus*; anterior wings more elongated, with their costal margin abruptly curved near the base.

Above, the four spots in the cell of the upper wings linear, of equal width, not increasing in thickness from the base outwards, as in *P. Eurypylus*; the macular band narrower, nearly white on the lower wings; abdomen and abdominal margin pure white.

Beneath, the red anal spot is not produced upwards along the abdominal margin, the pearly spots have a distinct dusky border, owing to their exceeding in size those on the upper surface. Expanse of wings 4½ inches.

Hab. Celebes (*Wall.*).

Remarks.—This is a powerful species of very rapid flight, and difficult to capture. It comes about muddy places in the villages of South Celebes, and is also found abundantly at pools in the half-dry mountain-streams. I consider it quite distinct from all the allied forms.

117. PAPILIO ÆGISTUS, Linnæus.

P. Ægistus, L.; Cram. Pap. Ex. t. 241. f. C, D; Boisd. Sp. Gén. Lép. p. 241.

Hab. Ceram, Gilolo, Batchian, Aru Islands (*Wall.*)

118. PAPILIO AGAMEMNON, Linnæus.

P. Agamemnon, L.; Cram. Pap. Ex. t. 106. f. C, D; Boisd. Sp. Gén. Lép. p. 240.

This species presents numerous slight modifications of form and marking, which seem hardly prominent enough to characterize as species, though tolerably constant in each locality. Type tailed.

Hab. India, Manilla.

Local form *a.* Tail shorter; wings rather pointed.

Hab. Timor, Flores (*Wall.*).

Local form *b.* Tail as in the last; two outer rows of spots on the lower wings absent.

Hab. Ké Island (*Wall.*).

Local form *c.* Size small; tail very short.

Hab. Malacca, Sumatra, Borneo, Java (*Wall.*).

Local form *d.* Wings much elongated, abruptly curved near the base; tail very short; size large.

Hab. Celebes (*Wall.*).

Local form *e*. Broader and less sinuated wings, body large, tail very short.
Hab. Ceram, Bouru, Batchian (*Wall.*).

Local form *f*. Form of *e*; tail reduced to a tooth; markings and spots well defined, rounded.
Hab. New Guinea, Aru Islands, Waigiou (*Wall.*).

119. PAPILIO RAMA, Felder.

P. Rama, Feld. Lep. Nov. Mal. p. 1.   *P. Arycles*, Boisd. Sp. Gén. Lép. p. 231 ?
Hab. Malacca, Sumatra (*Wall.*).

Remarks.—I have little doubt but this is the *P. Arycles* of Boisduval. His description, however, does not mention the distinctive character of the four large spots only in the discoidal cell; I have therefore used Dr. Felder's name.

LEPTOCIRCUS, Swainson.

This small but interesting genus differs somewhat from *Papilio* in the neuration of the wings, but is best distinguished by the longitudinal fold and great elongation of the hind wings. The species frequent water, often settling on the edges of rills, or hovering over pools and rivulets in the sunshine. The few species known are all very closely allied, and might with equal propriety have been considered as local forms of one species. Three have been already described, and I have therefore thought it better to add one more, than to attempt to reduce those which have been generally accepted as species to a lower rank.

120. LEPTOCIRCUS MEGES, Zinken.

P. Meges, Zink. Beitr. Ins. Java, p. 161, tab. 15. f. 8.   *Leptocircus Curius*, Sw. Zool. Ill. pl. 106 ; Boisd. Sp. Gén. pl. 7. f. 1, pl. 17. f. 3, p. 381.
Hab. Java, Malacca (*Wall.*).

121. LEPTOCIRCUS CURTIUS, n. s.

Larger than *L. Meges*; outer black margin broader, and apical nervures thicker; bluish band much narrower, of equal width on both wings, straight, abruptly narrow where it crosses the discoidal cell of the fore wings, and rounded at the inner margin so as to form a small notch at the junction of the fore and hind wings.

Under side with the band bluish silvery; the three small bands on the anal margin differing from those on *L. Curius* and *L. Meges*, the first being transverse, and not produced obliquely to join the vertical band, the second small and nearly obsolete, the third at the anal angle transverse, very little curved, and sharply defined.

Body beneath and base of all the wings greenish ashy.   Expanse of wings $1\frac{9}{10}$-2 inches. Length, head to tip of tail $2\frac{6}{10}$ inches.
Hab. Celebes (*Wall.*).

122. LEPTOCIRCUS DECIUS, Felder.

L. Decius, Feld. Lep. Nov. Philipp. p. 13.   *L. Curius*, G. R. Gray, List of Pap. in Brit. Mus.
Hab. Philippine Islands.

123. LEPTOCIRCUS CURIUS, Fabricius.

L. Curius, Fab. Ent. Syst. iii. 1. p. 28; Doubleday, Zoologist, 1843, p. 111; Gen. of Diurnal Lep. pl. 4*. f. 1; Don. Ins. Ind. pl. 47. f. 1.

Hab. Java (*Wall.*), North India.

NOTE.

In referring to the species described by Dr. Felder, I have quoted from papers which he has sent me, with distinct titles and separate paging, but which were all first published in the 'Wiener Entomologischen Monatschrift,' viz. "Lepidopterologische Fragmente" (quoted as "Lep. Fragm."), published at intervals from June 1859 to August 1860, "Lepidoptera Nova Malayica" (quoted as "Lep. Nov. Mal."), published in 1860, and "Lepidoptera Nova a Dr. Carolo Semper in insulis Philippinis collecta" (quoted as "Lep. Nov. Philipp."), published in 1861. It is to be regretted that the titles and paging of these separate papers were not made to correspond with the original publication, so as to have made a more exact reference possible.

I have also quoted Zinken's 'Beitrag zur Insecten-Fauna von Java' separated from the 'Nova Acta Acad. Nat. Curios.'; but in this case the pages and the numbering of the plates have been preserved as in the original work.

EXPLANATION OF THE PLATES.

PLATE I.

Represents the various forms of *Papilio Memnon* (see pages 6 and 46). N.B. The left side of each figure shows the upper surface, and the right side the under surface of the same insect.

Fig. 1. A male, from Borneo (a slight local variety).

Fig. 2. A female, from Java (a variety like *P. Agenor*, Cr.).

Fig. 3. A female, from Sumatra (a variety near *P. Anceus*, Cr.). The last two are varieties of the 1st dimorphic form of female in this species.

Fig. 4. A female, from Java (*P. Achates*, Cr.). The 2nd dimorphic form of female of *Papilio Memnon*.

PLATE II.

Represents the various forms of *Papilio Pammon* (figs. 1, 3, 5, and 6) and *P. Theseus* (figs. 2, 4, and 7). (See pages 6, 7, 51, 52, and 53.) N.B. The left side of each figure shows the upper surface, and the right side the under surface of the same insect.

Fig. 1. *Papilio Pammon*; a male, from Malacca.

Fig. 3. The first form of female, closely resembling the male, from India.

Fig. 5. The second form of female (*P. Polytes*, L.), from Singapore. This is the most common and widely distributed form of female, occurring everywhere with the male.

Fig. 6. The third form of female (*P. Romulus*, Cr.), from India.

Fig. 2. *Papilio Theseus*, the first form of female, almost exactly resembling the male, from Timor. This form is very rare.
Fig. 4. The second form of female, from Timor.
Fig. 5. The third form of female (*P. Theseus*, Cr.), from Sumatra. The second and third forms of female seem about equally plentiful, but are generally confined to separate islands. A fourth form of female (*P. Melanides*, De Haan) would have been figured, but could not be brought on to the plate. (See pages 7 and 53.)

Plate III.

Represents the various forms of *Papilio Ormenus* (see pages 8, 55, and 56). N.B. The left side of each figure shows the upper surface, and the right side the under surface of the same insect.
Fig. 2. A male, from the island of Goram.
Fig. 1. The first form of female, from Waigiou.
Fig. 3. The second form of female, from Waigiou.
Fig. 4. The third form of female (*P. Amanga*, Bd.), from the island of Goram.

Plate IV.

Represents two species allied to *Papilio Ormenus*, but whose females are not dimorphic (see pages 57 and 58). N.B. The left side of each figure shows the upper surface, and the right side the under surface of the same insect.
Fig. 1. A female of *Papilio Adrastus*, peculiar to the island of Banda (see page 57).
Fig. 3. *Papilio Tydeus*; a male, from Batchian.
Fig. 2. The female of *Papilio Tydeus*, exhibiting a single permanent form confined to a small group of islands (Batchian and Gilolo), intermediate between the two forms of *Papilio Ormenus* ♀ which are represented on Plate III. figs. 3 and 4.

Plate V.

Represents several new species of *Papilio*, illustrating "local variation." N.B. The right side of each figure shows the upper surface, and the left side the under surface of the same insect.
Fig. 1. The male of *Papilio Noctis*, from Borneo (see page 41). The female was figured by Mr. Hewitson in the 'Proceedings of the Zoological Society of London,' 1859, plate 66. fig. 5.
Fig. 2. *Papilio Leodamus*, male, from Mysol (see page 42).
Fig. 3. *Papilio Hecuba*, male, from Celebes (see pages 16 and 50).
Fig. 4. *Papilio Pertinax*, male, from Celebes (see page 49).
Fig. 5. *Papilio Albinus*, male, from New Guinea (see page 49).

Plate VI.

Represents four species not before figured, belonging to the most brilliantly coloured group of Eastern Papilios, and illustrating local modifications of form. N.B. The right side shows the upper surface, and the left side the under surface of the same insect.
Fig. 1. *Papilio Pericles*, male, from Timor (see page 45).
Fig. 2. *Papilio Macedon*, male, from Celebes (see page 45). This species exhibits in a marked manner the strongly arched wings characteristic of those from Celebes, as contrasted with those represented at figs. 1 and 3, from other islands (see pages 16, 17 and 18).
Fig. 3. *Papilio Philippus*, female, from Ceram (see page 45).
Fig. 4. *Papilio Blumei*, male, from the north of Celebes (see page 46). This also exhibits the arched wing, as compared with its ally from the Moluccas (fig. 3).

PLATE VII.

Represents six remarkable species of *Papilio* not before figured. N.B. Except in fig. 1, the right side shows the upper surface, and the left side the under surface of the same insect.

Fig. 1. *Papilio Thule*, male. The upper surfaces of two varieties or local forms of this species are figured. The right side represents the form found in New Guinea, the left side that obtained in Wargiou. It resembles *Danais sobrina*, Bd., which inhabits the same countries, and varies in a somewhat similar manner (see pages 20 and 64).

Fig. 3. *Papilio Ænigma*, male, from Sumatra (see page 60). This species was named as above, from its puzzling resemblance to *Papilio Paradoxa*, which is found in the same districts. Both species appear to mimic *Euplœa Midamus* (see page 20).

Fig. 2. *Papilio Miletus*, male, from Celebes (see page 65). This species and the next exhibit in a striking manner the abruptly curved wing peculiar to Celebes. Figs. 5 and 6 represent species almost equally remarkable in this respect.

Fig. 4. *Papilio Telephus*, male, from Celebes (see page 67).
Fig. 5. *Papilio Androcles*, male, from Celebes (see page 63).
Fig. 6. *Papilio Gigon*, female, from Celebes (see page 59).

PLATE VIII.

Illustrates, by comparative outlines of the anterior wings, the local modification of form in the Papilios of Celebes as compared with those of the surrounding islands. In each pair of outlines, the upper one represents a species peculiar to Celebes, while the one beneath it shows the most closely allied species or variety from any of the surrounding islands. (For details, see page 16.) The following are the names of the species:—

Fig. 1. *Papilio Gigon*, from Celebes; *P. Demolion*, from Java.
Fig. 2. *Papilio Macedon*, from Celebes; *P. Peranthus*, from Java.
Fig. 3. *Papilio Androcles*, from Celebes; *P. Antiphates*, from Borneo.
Fig. 4. *Papilio Telephus*, from Celebes; *P. Jason*, from Sumatra.
Fig. 5. *Papilio Miletus*, from Celebes; *P. Sarpedon*, from Java.
Fig. 6. *Papilio Agamemnon*, var., from Celebes; *P. Agamemnon*, var., from Sumatra.

PLATE VI.

Represents four species not before figured, belonging to the most brilliantly coloured group of Eastern Papilios, and illustrating local modifications of form. N.B. The right side shows the upper surface, and the left side the under surface of the same insect.

Fig. 1. *Papilio Pericles,* male, from Timor (see page 45).

Fig. 2. *Papilio Macedon,* male, from Celebes (see page 45). This species exhibits in a marked manner the strongly arched wings characteristic of those from Celebes, as contrasted with those represented at figs. 1 and 3, from other islands (see pages 16, 17, and 18).

Fig. 3. *Papilio Philippus,* female, from Ceram (see page 45).

Fig. 4. *Papilio Blumei,* male, from the north of Celebes (see page 46). This also exhibits the arched wing, as compared with its ally from the Moluccas (fig. 3).

Wallace, A. R., "On the phenomena of variation and geographical distribution as illustrated by the Papilionidae of the Malayan region," The Transactions of the Linnean Society of London, V. 25, 1866, p. 1–71.

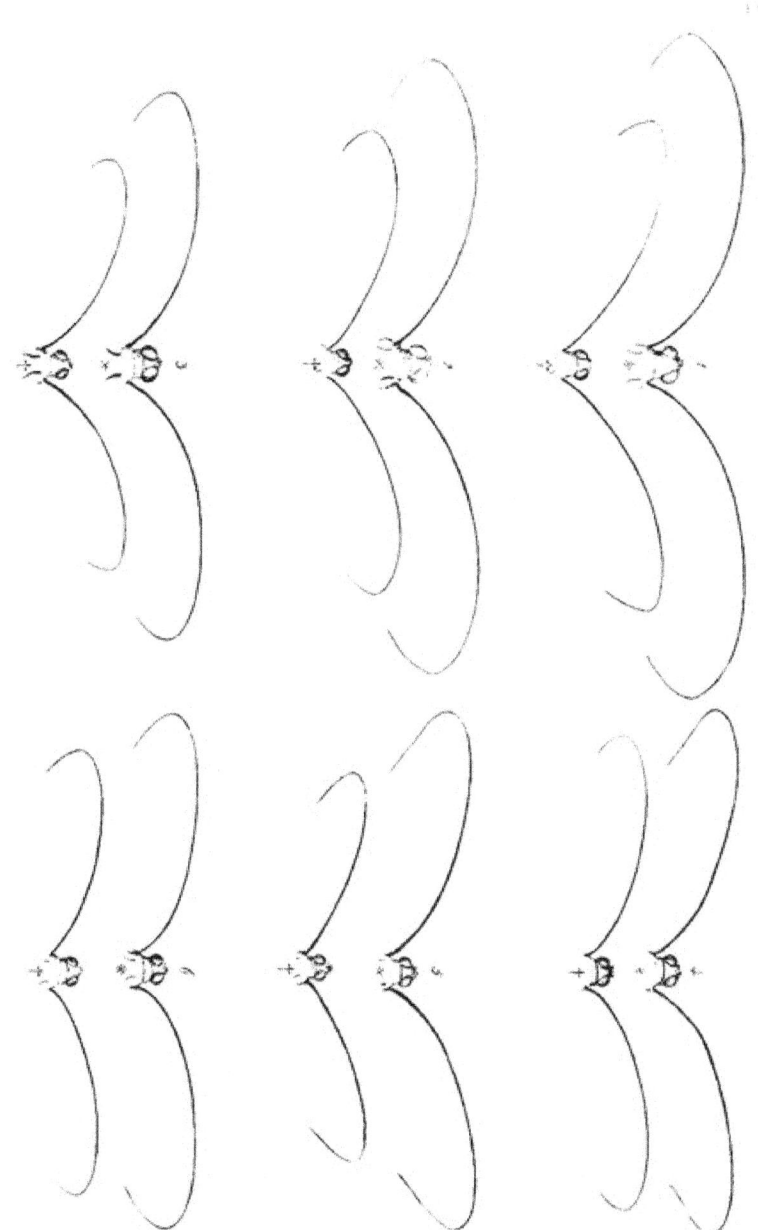

www.ingramcontent.com/pod-product-compliance
Lightning Source LLC
Chambersburg PA
CBHW020300090426
42735CB00009B/1154